Good Enough to Eat?
Next Generation GM Crops

Good Enough to Eat?
Next Generation GM Crops

Ian D. Godwin
The University of Queensland, St Lucia, Australia
Email: i.godwin@uq.edu.au

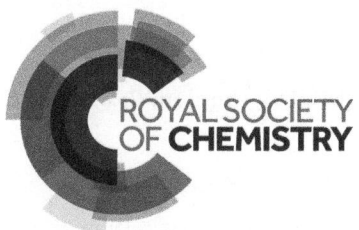

ROYAL SOCIETY
OF CHEMISTRY

Print ISBN: 978-1-78801-085-6
EPUB ISBN: 978-1-78801-681-0

A catalogue record for this book is available from the British Library

The Royal Society of Chemistry is a charity, registered in England and Wales, Number 207890, and a company incorporated in England by Royal Charter (Registered No. RC000524), registered office: Burlington House, Piccadilly, London W1J 0BA, UK, Telephone: +44 (0) 20 7437 8656.

Visit our website at www.rsc.org/books

Printed in the United Kingdom by CPI Group (UK) Ltd, Croydon, CR0 4YY, UK

Foreword

This is a book that will make you laugh, and weep, and even be angry – more or less in equal proportions. And this is surprising because it is about genetics and plant and animal breeding. But actually, not so surprising because the popular dialogue about modern genetic breeding is characterised by these emotions.

Ian Godwin is a first-class scientist, a professor of plant molecular genetics at the University of Queensland in Australia, with 30 years of experience in biotechnology research. He works on a wide variety of crops, ranging from sorghum, wheat and barley, to beans and taro, creating knowledge of their molecular structure, their genetics, how they grow and how they can be engineered to give us answers to one of the world's most important challenges.

Despite the progress we have made over recent decades we are still a long way from achieving food security for all – for every man, woman and especially for every child – on this planet. The challenges are considerable – from pests and diseases, eroded soils, to lack of water and the impact of climate change. On top of this there is a highly vociferous and unprincipled opposition from some sectors of the public, who choose to ignore facts and realities.

The cast includes scientists, politicians, company PR people, NGOs big and small and even pantomime characters. Ian has travelled and talked to many of the key actors and uses their

Good Enough to Eat? Next Generation GM Crops
By Ian D. Godwin
© Ian D. Godwin 2019
Published by the Royal Society of Chemistry, www.rsc.org

words to create a lively dialogue. Some of the stories are well known, others less so, but all are told in a way that illuminates and often challenges. Perhaps the most disturbing was the Great Petunia Massacre.

Ian has the enviable gift of taking complex biological, and indeed socio-political, processes and making them not only readily understandable but also amenable to thoughtfulness. One of the most impressive chapters, entitled *New Kid in Town*, is about the modern science of gene editing, and even I, as a non-molecular biologist, at last think I understand how it is accomplished and begin to appreciate just what a revolution it promises, one that will soon make GM seem old hat. A recent public meeting at the Royal Society in London had hundreds of young people lining up around the block to listen to one of the world experts in gene editing. We live in exciting times.

The book does not have to be read from beginning to end. Each chapter has its own distinctive stories – whether about cotton in Burkina Faso, or Golden Rice or Monarch butterflies or late blight. The reader can choose any chapter, perhaps on the basis of an intriguing title, and start from there, working forward or back or skipping to another. Enjoy!

Sir Gordon Conway, KCMG, FRS, HonFREng
Professor of International Development
Imperial College
London

Preface

First and foremost I would like to thank Monsanto for not funding this book and for also funding a grand total of zero percent of my research. Ha, ha! You wish! Truth be told, sometimes I sort of do too. Undertaking agricultural biotechnology research remains a constant struggle to find sources to fund all of our good ideas, from tube to cell to Petri dish to plant to crop. I think it was Ernest Rutherford who said "We've got no money, so we've got to think." Luxury! Well, maybe that works for a theoretical physicist, but not for a mere agricultural scientist like me. No money = no research = no job.

Yet that is the situation for many agricultural biotechnologists in many parts of the world. Imagine working in a field of research where you seek to understand and harness plant, animal and microbial genetics and biology for the benefit of humankind. Yet you can't even go to the local pub and tell people what you do without running the risk of being abused for such huge crimes as putting "gay genes" in corn, causing cancer *via* soybeans, premature puberty with canola and forcing most farmers in India to commit suicide because they dared to grow cotton that had a natural gene for insect resistance. This is the situation that many agricultural biotechnologists find themselves in on a regular basis. Of course, none of these things happened, but like the 1969 landing on the moon being filmed

Good Enough to Eat? Next Generation GM Crops
By Ian D. Godwin
© Ian D. Godwin 2019
Published by the Royal Society of Chemistry, www.rsc.org

in a Hollywood basement, they have gained currency as conspiracy theories.

When approached by the Royal Society for Chemistry to write a book about GM foods and crops, I instantly channelled that great turf scientist, John McEnroe. He had an innate skill in being able to distinguish between plants (turf grass) and chemicals ("I saw chalk"). Notwithstanding the fact that McEnroe has, very likely, an excellent understanding that the grass courts of Wimbledon are a mixture of 70% perennial ryegrass (*Lolium perenne*) and 30% creeping red fescue (*Festuca rubra*), he may even know that the "chalk" is actually a bunch of chemicals, like titanium oxide to show those lines brighter than bright on television. However, I thought "YOU CANNOT BE SERIOUS". Me? Write a book?

The alternatives were easy. I could just stay bunkered down in my lab continuing to fight the good fight *via* my research papers, some of which have been cited by hundreds of other scientists. Perhaps I could occasionally put my head above the parapet and write 800 words or so on GM crops for media outlets like *The Conversation*. Or I could take some advice and refer to Rutherford again: "A theory that you can't explain to a bartender is probably no damn good."

The prospect of walking to my local, the historic Regatta on the picturesque Brisbane River, and regaling the person behind the bar with the benefits of GM crops had little appeal. Have you ever worked behind a bar? All sorts of nut-jobs will approach the bar for meaningful conversations, usually fuelled by the fermented products of cereals, tubers and fruits. "Gidday love, did you know if you use RNAi to downregulate an endosperm-specific protein foldase gene in sorghum you can get more seed storage proteins and larger seeds? When you get off work, wanna come back to my lab and see the size of my panicles?" Not a conversation anybody should be subjected to.

So I decided that it may well be more effective, and safer, to write a book instead. My darling wife is very happy for me to show her the size of my panicles, but then again, she likes me and indulgently listens to most of my stories. My wonderful children also kindly listen to my stories too, and now I have a little grandson, the next victim in line. You, however, are busy, and reading this book will have to compete with the time you may have fruitfully spent doing other things. Of necessity,

I enlisted the help of other storytellers. Storytellers who are also food scientists, biotechnologists, geneticists, agricultural scientists, farmers, chefs, social scientists and even activists with an interest in the way we view food and agriculture. These storytellers kindly shared their wonderful stories with me, and I have done my utmost to accurately reproduce their science, their anecdotes and their opinions. Any errors of fact are my own.

There are many people I would like to thank. Most of all I thank my wife and best friend, Melissa Glendenning, for always encouraging me to write, even taking over the dining table with all my disorganisation. I also thank my University of Queensland colleagues, Karen Massel and Brad Campbell, for their thoughtful and tactfully critical feedback on each chapter of this book. I would particularly like to acknowledge UQ for allowing me to take a six-month Special Studies Program in Europe to work on this book. I would like to express my deep gratitude to Professor Rod Snowdon of Justus Liebig University, Giessen, who let me stay in his house and welcomed me into his family for three months. I would also like to express my great appreciation to Professor Birger Lindberg Møller of the University of Copenhagen. Birger not only hosted me in beautiful Copenhagen, but was instrumental in securing me an historic Danish Nationalbanken apartment in Nyhavn, in the apartment block where Hans Christian Andersen spent the final years of his life.

Rest assured, however, this book is not a fairy tale. Despite the semblance of being a collection of anecdotes, this book predominantly depicts evidence-based science. Some parts of this book may infuriate you, as they did me while I wrote it. I hope you enjoy the time you invest in these pages, and that some of these storytellers induce you to re-evaluate what you think you know about what food is, indeed, good enough to eat.

Ian Godwin

Acknowledgements

So many people have been wonderfully helpful and encouraging to me in bringing this book to life. I cannot really ever thank Joan, my mum, and Doug, my dad, who are the reason I became a scientist. As well as the usual gifts of cricket bats and bicycles, it was really those special things like a microscope, a chemistry set (with matches and a spirit burner!) and books like *Tell Me Why* by Arkady Leokum (1965) that set me on the path. Yet in truth, it was when my mum told the sports-mad me that I could not take Physical Education as a senior subject that I had to choose Biology instead (yawn). After a semester of genetics I found a new purpose in a subject that still thrills and surprises me 40 years later.

I cannot thank all the wonderful people who gave freely of their time in being able to tell their stories and add their knowledge and experience to this book. I hope you agree with me that there are some very inspirational stories here. They are all very different people, in some cases absolute characters, who really make being a scientist the absolutely best job in the world. Unless you have an eccentric way of reading a book, you are about to meet these people and through this book hear something of their stories. Without going through the list exhaustively, there are a few people who I must give extra thanks to. When I first embarked on the interviews, and wasn't sure that I had any idea what I was doing, Klaus Amman and Peggy Lemaux

Good Enough to Eat? Next Generation GM Crops
By Ian D. Godwin
© Ian D. Godwin 2019
Published by the Royal Society of Chemistry, www.rsc.org

gave me so much information and encouragement that I felt I might actually be able to make this happen. Then after speaking with infectiously enthusiastic advocates for biotech and story-tellers like Alison van Eenennaam (after a launch of the movie *Food Evolution* at a conference in San Diego), and Luis Hererra-Estrella it just became easier.

I had a lot of fun working on my book while at Justus Liebig University in Giessen with Rod Snowdon, who introduced me to Urs Niggli. Urs kindly hosted me at the Research Institute for Organic Agriculture in Switzerland and gave me many insights into organic agriculture. Thanks also to Birger Lindberg Møller, who hosted me at Copenhagen University for three months and introduced me to Ross Cloney – great chats, great tatts. Cami Ryan was also extremely generous with her time, and put me in touch with Stuart Smyth. I am greatly indebted to Sir Gordon Conway, who kindly agreed to write the Foreword, and is a man who has inspired young agricultural scientists all over the world.

Thanks also go to my University of Queensland colleagues who generously gave of their time, Heather Smyth, Liz Aitken, Jimmy Botella and André Drenth, who was a fantastic source of information when it came to potato blight, potatoes in general, and also showed that he was a great chapter editor. Members of my lab, Brad Campbell and Karen Massel, were fantastic at reading all my drafts and making suggestions to improve each and every chapter – and all in return for beer! I think I still owe them many beers for a long time to come. Lara-Simone Pretorius from the lab upstairs also provided some excellent illustrations, all while finalising her PhD. Nicola MacKay, a member of Australia's largest banana-growing family, was extremely generous with her time. The future of Australian agriculture is in great hands with Nicola (full disclosure – she is also a UQ student). I also thank my many other UQ friends who constantly asked "How's the book going?" because it always helped me to focus on what was important and ignore some of the administrivia piling up at UQ.

The wonderful people at the Royal Society for Chemistry have been the most understanding publishers I could have imagined. Thanks to Rowan Frame, the editor who talked me into all this, and the wonderfully patient Drew Gwilliams and Katie Morrey, who have helped me every step along the way.

Finally I would like to thank my wonderful wife, Melissa Glendenning. Without her constant love and encouragement, I sincerely doubt I would have made it to the end of this book. She was also cheerfully tolerant of the way I seemed to somehow take over the dining table for long periods of time with paper, paper and more paper! TV dinners are a thing of the past ... until my next book.

Dedication

For Harry, Melissa and Angus
The future is in great hands with you

Good Enough to Eat? Next Generation GM Crops
By Ian D. Godwin
© Ian D. Godwin 2019
Published by the Royal Society of Chemistry, www.rsc.org

Contents

Chapter 1
Food, Glorious Food   **1**

References   19

Chapter 2
A Kind of Magic   **20**

References   51

Chapter 3
Revolution   **52**

     3.1   Herbicide Resistance – Roundup Ready   56
     3.2   Insect Resistance – Bt Cotton and Maize   62
     3.3   Bt Maize   68
     3.4   Disease Resistance – the Virus-resistant Rainbow Papaya   68
   Bibliography   73
   Reference   73

Chapter 4
Chemical Heart   **74**

   Bibliography   95
   References   95

Good Enough to Eat? Next Generation GM Crops
By Ian D. Godwin
© Ian D. Godwin 2019
Published by the Royal Society of Chemistry, www.rsc.org

Chapter 5
Wide Open Spaces   **96**

    5.1   Soybean and Roundup Ready   108
    5.2   GM Maize in the USA and Africa   115
    5.3   GM "Rainbow" Virus-resistant Papaya   119
   References   119

Chapter 6
Bad Moon Rising   **121**

    6.1   Highly Nutritious Soybean   138
    6.2   Kinki Pigs   143
    6.3   StarLink Maize   149
   References   155

Chapter 7
Paint It Black   **157**

    7.1   GM Foods are Toxic   173
    7.2   Farmer Suicides   175
    7.3   Bt Maize Killed Monarch Butterflies   179
    7.4   GMO = Pesticides   185
    7.5   GMO is Not Organic   186
   References   187

Chapter 8
Not Ready to Make Nice   **189**

   Further Reading   211
   References   211

Chapter 9
O Fortuna!   **213**

   References   241

Chapter 10
New Kid in Town   **243**

   References   268

Chapter 11
For a Better Day 270

   Bibliography 297
   References 298

Subject Index 300

CHAPTER 1

Food, Glorious Food

According to Franz Kafka:

"So long as you have food in your mouth, you have solved all questions for the time being."

Well, is that the case? I remain unconvinced. What if the food contains gluten? Or peanuts? Or horse? Or Brussels sprouts? Or what if it's genetically modified? Is it good enough to eat?

Just reading this has conjured up all sorts of emotions for you. Some of you are saying, "Of course it's good enough to eat". My wife is saying, "Why did you bring me to a place that has horse on the menu?" Some of you are reaching for an EpiPen.

Food. It's complicated. It's not just for survival. It brings people together. It divides people. It's for health (good or bad). It's for ceremonies. It's for bingeing. It's for dieting. It's for boasting about. It's for complaining about. It's delicious. It's disgusting – how can you eat that? Here comes an aeroplane. No, don't put it in your hair. And disturbingly, for many people it's just not available in decent quantity or quality.

So, while this book is about how new genetics in the 20th and 21st centuries has revolutionised food production, it's impossible to have a meaningful dialogue without acknowledging and exploring the complex relationship all of us have with food. Many of us have little or no personal experience or knowledge

Good Enough to Eat? Next Generation GM Crops
By Ian D. Godwin
© Ian D. Godwin 2019
Published by the Royal Society of Chemistry, www.rsc.org

about where our food comes from. However, many others know exactly the where and how, because some people are subsistence farmers who have to grow, hunt or gather everything they eat. I'm betting you are not a subsistence farmer, because if you were you would be unlikely to have time to read this book. You'd be up at dawn fetching water and firewood, and getting the kids to milk the scrawny cow tethered outside while your husband went to look for the goat that disappeared overnight.

No, you most likely got your milk from a plastic bottle that was shipped from 100 to 2000 km away to have with your coffee that was grown in Africa or South America. I'm doing the same. So that's OK.

So in this chapter, we are going to have a conversation about food and some of our complex and wonderful relationships with it. We need to start in this place, because many of our attitudes to new technologies in food are guided by our relationship with food. We all want to eat tasty, nutritious and sustainable food (and more about sustainability later), and once you get to be 50 something, you'll probably have to eat less of it or end up buying a bigger pair of trousers. But that's enough about me, let's start the journey. And all good journeys require travel, so we'll do a lot of that in this book. Let's start in France. Not just because it's a great place for food, but because some very important Australian history took place in France. We are not going to the sunny south where the miracle of the Mediterranean diet will keep you alive for 200 years without ever having to buy a larger dress size. No, we are going to the north, where it's a bit colder, a bit greyer, and they serve cream and butter with almost everything. And a century ago it was full of 295 000 young Australian men.

We're "on the Somme", 2017. The soggy dawn of an early spring day is dreary and grey. Standing in a war cemetery in Dranoutre, looking at the grave of a great-uncle who was killed almost 100 years ago does little to improve the mood. Uncle Jack Bond, an Australian artillery man, was mortally wounded by shrapnel on 30 March 1918.

> *I was next to Bond when he was killed. He was hit in the head and only lived a few minutes. This was at Dranoutre, the 47th Battalion was resting at the time. Bond came from Sandgate Park, Queensland.*
> Witness statement from Gunner R.S. Colbourne, Glebe, Sydney, 1918

Unbeknownst to Gunner Bond, only about 10 km up the road was my grandfather Vincent George Flanagan, who after the war married Jack's little sister, Olive Bond. Vince was a railwayman all his working life. In 1916 he joined up in Queensland, and disembarked from Melbourne as a trained artilleryman with the 3rd Field Artillery Brigade. Vince rarely talked about the war, but he did once tell me the artillery were called the "drop 'em shorts" and it was not a term of endearment.

Vince never served as an artilleryman, because once in Britain, he was quickly redeployed into a new Engineer's Unit, known originally as the Victorian Railway Unit, later the 15th Light Railway Operating Company, and "taken on strength" in Poperinghe, in October 1917, just before the Australian attack on Passchendaele with the Canadians. A month later, the Australians withdrew having suffered 38 093 casualties.

My grandfather then endured a winter with occasional shelling and bombing from what I can glean from the CO's official diary. On 30 December 1917, he was recommended for a Military Medal, and apart from "on the Somme" nobody in the family was ever privy to the circumstances. Being an enlisted man there was no official citation. When I was 10, my granny died and Vince came to live with us. Being a boy, I was naturally curious about his war experience – asking insightful and sensitive questions that only a 10-year-old can ask, like "What did you do in the war?" or "Did you kill any Germans?" The only answer I ever received from my skilful line of questioning was, "I had to shoot a lot of poor bloody horses who got stuck in the mud". He was a great horse lover, and did spend a significant proportion of his weekends (and wages) at the races.

By now, you the reader will be wondering whether this book has been bound in the wrong cover. Apologies if you found little of this particularly informative or relevant, but it does serve as a lengthy preamble to the only other thing I was ever able to find out about my grandfather's war experience. One of those mornings at breakfast, being a helpful, polite young fellow schooled in perfect table manners by the steely determination of both mum (Vince's daughter) and dad, I asked grandfather if he would like jam on his toast.

"What sort is it?"

"Plum. . . .."

"I vowed if I ever survived the bloody war I would never eat bloody plum jam again in my life".

Now I don't think post-traumatic stress disorder had been invented in the early 1970s, and even if it was, this 10-year-old boy was blissfully unaware of its existence. With the benefit of hindsight and maturity, I suppose this was one of my grandfather's few outward symptoms of PTSD. In my 10-year-old "Boys' Own" excitement, I pictured Vince ensconced in a muddy shell hole with nothing to eat but a can of plum jam for days on end, until he was relieved with a can of Fray Bentos bully beef. Or I envisaged a scenario where he and his mates captured some Germany army issue "marmalade". They toasted up some baguettes they had "foraged" from a local farmhouse, but on opening the cans found the dastardly Hun had played a filthy trick. No orange had been harmed in the making of this marmalade (generic German for jam), and all the tins contained plum jam! Crikey! But I'm pretty sure that never actually happened either and I most certainly appropriated that memory from an old episode of "Dad's Army".

Nevertheless, as Napoleon, or the Duke of Wellington, or maybe even Russell Crowe said in some movie, an army marches on its stomach. That's not really the point – and the point is that food is not just something we eat to stay alive, it is a core part of our emotions, our physical and mental health and wellbeing, and indeed, our very identity.

Fortunately for my grandfather (and for my yet-to-be-conceived mother and, ultimately, me), the 15th Australian Light Railway Operating Company was involved in preparations to defend against the much-anticipated German "big push". Doubtless he didn't feel too fortunate at the time, and throughout early March 1918, they were shelled by day and bombed by night. A diary entry by Lance Corporal William Lycett of 8 March:

After dinner went down and started unloading 6-inch howitzer ramps used for carrying guns on light railway trucks, very heavy work, lot of our chaps getting sick leaving us short-handed.

Vince was one of the chaps who got "sick". In his service record it simply states "9th March 1918 hernia – evacuated to Edgbaston, England". He survived the operation, the convalescence and even a few days in the clink for "failing to salute a British Officer at the Changing of the Guard, Buckingham

Palace" and returned to Australia to eventually marry Ollie and have a daughter, my wonderful mum, Joan.

So a hundred years after all this history, here I am living in the 21st century, and I am able to do something my children take for granted, but Vince never lived to experience – I Googled "plum jam WWI" and gained some insight into the plum jam story. Suffice to say that it is highly unlikely that Vince was the only Allied serviceman who swore off plum jam during that terrible time on the Somme, or anywhere else on the Western Front.

A company called Ticklers in Grimsby in the north of England had the government contract to supply plum jam to the British and Dominion troops at the front. And in case you didn't know, Australia was a Dominion of the British Empire back then, along with Canada, New Zealand and South Africa (as well as Newfoundland and the Irish Free State). More about the jam itself later, but the tins themselves played a key role in trench warfare. Early in the war the Germans had a good supply of "stick bombs", whereas the Allied troops suffered a critical shortage of the Mills bomb, what we would today call a grenade. So, in Gallipoli and later on the Western Front, the Aussie, Canadian and British troops improvised their own bombs by stuffing empty Tickler's jam tins with nails, screws and gunpowder. The "Tickler's Bomb" became a standard.

> *Tickler's jam, Tickler's jam*
> *How I love Tom Tickler's jam*
> *Plum and apple in one pound pots*
> *Sent from England in ten ton lots*
>
> *Every night when I'm asleep*
> *I dream that I am*
> *Bombing the poor old German's trench*
> *With Tommy Tickler's jam.*

The tins were sought after, whereas the jam, sadly, was not. Indeed, there is ample evidence to suggest that the involvement of plums and apples was minimal when it came to the manufacture of Tickler's Plum and Apple. Many contemporary witnesses reported that the factory in Grimsby received frequent shipments of turnips and swedes. One of Bruce Bairnsfather's iconic WWI postcards depicts a down in the mouth soldier eliciting The Eternal Question "When the 'ell is it going to be strawberry?". Plum and apple jam became a symbol in the trenches of getting second best or "short shrift".

Another song from the soldiers in the trenches was a lament on getting second best, with the suspicion that the good stuff was kept by the Army Service Corps (ASC):

Plum and apple
Apple and plum
Plum and apple
There's always some

The ASC gets strawberry jam
And lashings of rum
But we poor blokes
We only get –
Apple and Plum

Perhaps now more than ever in human history, we identify (with all our other cultural mores of "belonging") *via* what we eat – or sometimes more importantly, what we don't. In the Western world my generation and those following have forgotten that only a couple of generations ago most of our antecedents had a great longing for a wider variety of food, and MORE OF IT.

My parents were born into 1930s Australia (the Great Depression) and 1940s, where wartime rationing continued right up until 1950 for tea and butter. My dear and funny dad, Doug, still talks about having bread and dripping (the fat that has melted and dripped from roasting meat) for Sunday dinner. Seeing the frequently unfettered consumption of our boys today, the fact that my dad and his mates survived at all is something to contemplate. We now live in a time of plenty, and with that plentiful bounty comes the ability to choose.

Although it wasn't always like that in my family, we had to finish everything that was on our plate. Sometimes, like when we went to Aunty Ruby's for Sunday lunch, we could have seconds. My strategy for survival was to "save the best for last", invariably the meat, usually starting with the green or yellow vegetable (10-year-old boy shudders). The horror of being confronted with such things as the humble choko (Figure 1.1) leaves a strong

Figure 1.1 *Sechium edule* (choko or chayote), a fruit deemed by some to be edible.
© Jiang Hongyan/Shutterstock.

childhood scar, not just in my mind, but for almost every kid in Australia growing up in the 1960s. To be fortunate enough to hate broccoli, or Brussels sprouts or turnip, was a luxury that only those of us not indelibly scarred by the ultimate food aversion therapy, the choko. Choko (even the name suggests a sense of foreboding doesn't it?) is the Australian descriptor for chayote (*Sechium edule*). *Edule* gives it away really. It means edible. Not *Sechium deliciosa* or *S. magnifica* or *S. epicure* – well any of those would have been in contravention of all advertising conventions, and failed the morality test even among taxonomists. The New Zealand Maori people survived on "edible" ferns, yet those times are the worst of times in Maori folklore. And the same goes for the folklore of any kid from Queensland in the 1960s, it's the choko. Well, that and the dreaded school milk. Every morning we had to drink our compulsory school milk, either at little lunch (morning break), or before the first lesson. In summer, after the milk had sat in the sun at around 28 °C for an hour or two, we were then faced with the indescribable joy of something formerly known as milk. It was an experience seared into the consciousness of every Queensland schoolkid until 1973. The Education Queensland website sums it up best:

The taste of the free school milk will remain vividly in the memory of school children from this era. The milk was never refrigerated and on a hot Queensland day, the taste it had

*acquired by "little lunch" could be sickening. Enjoyment was not
improved if you forgot to shake the bottle before opening and got
a mouthful of warm, sometimes lumpy cream.*

I remember it like it was yesterday.

We all have food (not necessarily culinary) experiences that are
linked strongly to memories of people and place. Some wonderful, some less so. A few years ago, my lovely wife, Milly, and I
were walking along the north coast of Spain, and happened
along the small fishing port of Getaria. Getaria was the birthplace of fashion designer, Cristobal Balenciaga. Yeah, OK, I'd
never heard of him either. Travel is about learning. Above the
port frontage is a bronze statue of a famous mariner, Juan
Sebastien Elcano. Who? Being a product of mid-20th century
British-centric education, I had it drummed into me that Captain
James Cook "discovered" Australia. While that does ever so
slightly overlook the claims that could be made by the indigen-
ous population, who obviously sat around gazing out to sea for
60 000 years waiting for a white guy with a funny accent to
"discover" the continent, there's also the various Dutch and
Portuguese and Melanesian and Indonesian seafarers who had
well and truly put the place on the map, such that the world
could rejoice at the discovery of one new food. The macadamia
nut, Australia's modest contribution to the world's food plants.

But back to Juan Sebastien Elcano, the guy immortalised in
bronze. Now if you learned at school that Ferdinand Magellan
was the first person to circumnavigate the globe, you too have
been misled. Magellan was killed in the Philippines, much in the
same way that Cook was in the Hawaiian Islands, although I
don't think Magellan was turned into a menu item that night. So
the truth is, Magellan didn't even get much further than halfway.
Perhaps we should have been taught that Magellan was the first
person to semi-circumnavigate the world. It was Basque seafarer,
Juan Sebastien Elcano, who led the only surviving boat with
merely 18 crew back to Spain to be forgotten as a footnote by
everyone, except the good people of Getaria. So there's two
"facts" I learned and successfully regurgitated at school during
exams, both skewered by reality. Don't believe everything on the
curriculum kids. The post-truth world has existed for quite some
time. But what made Getaria totally memorable wasn't the

bronze or the fashion. It was the smoky row of fish barbecues (Asador) along the waterfront. For me and Milly it was our first taste of monkfish and to this day, whenever I see the name Balenciaga, I think of barbecued monkfish. And whenever we have delicious fish, we have developed the habit of declaring it "almost as good as Getaria".

Not quite such an edifying experience, but decidedly memorable, was my first (and thankfully only) encounter with an Indonesian dish known as pecel lele. Pecel lele is a Javanese specialty – the description sounds innocuous enough, a deep-fried freshwater catfish served with chilli paste. Perhaps it was the presentation – a large bowl with a sea of red (that would be the chilli paste) from which was staring the ugliest, blackest fish I have ever encountered. When I say fish – it actually bore a striking resemblance to Jabba the Hut. Now, I regard myself as something of a foodie, and definitely adventurous, so I had a go. As we are in polite company, my brief tasting notes were muddy water, sulphur, something metallic – a bit like chewing aluminium foil, and Dante's Inferno with the hottest chillies I have ever encountered. Throughout the evening it was a gift that kept giving, and yes it was Dante's Inferno for the entire length of the alimentary canal. Science now tells us we have taste buds throughout our alimentary canal, and after pecel lele, I'm a believer.

Food is also an important part of our identity – personal, cultural and national – although my home country, Australia, is almost an exception here. We have embraced the cuisines of the world, and given that we started with a fairly bland rendition of British cuisine from our cultural heritage, diversity is a wonderful thing. Within 2 km of my house in suburban Brisbane I can go out and eat Italian, Greek, Chinese, Malay, Japanese, French, Persian, Mexican, US pizza and burgers, Vietnamese, Thai, Tibetan, Indian and Aussie pub food. What's Aussie pub food? It's all those others put together in one place, albeit not done particularly well or authentically. We don't have much to offer the world's cuisine, with the exception of Vegemite. If you've met an Australian you'll know it and probably have an opinion of it. Regardless of what you think or have heard, Vegemite on buttered toast is the best hangover food known to humanity.

Food is central to many cultures and nationalities. As Julia Child said, "People who love to eat are always the best people".

In many parts of east and south-east Asia, a common greeting is, "Have you eaten (yet) (rice)?". Food is so much part of us and our parlance that we "chew the fat", "bring home the bacon", become "the bread winner", have "a bun in the oven", are "as cool as a cucumber", "spill the beans" and say simple things are "as easy as pie" or a "piece of cake", show empathy by espousing "that's the way the cookie crumbles" or remonstrate approval with the epithet "the best thing since sliced bread".

Food is increasingly a personal identity handle for many. We've all met people who within the first 5 minutes have told us "Oh by the way I'm vegan" as if somehow this was as interesting as "Oh by the way I'm Donald Trump/a Nobel laureate/world record holder for hammer throw/your Father, Luke". Food has been ingrained into our culture since, well since culture started.

In my building at the University of Queensland, many of the Australians (academics, researchers and students alike) will sit at their desk and eat their lunch. I'm guilty of this most days too, there always seems so many things to catch up on that you can't contemplate taking 45 minutes off to eat lunch and "chew the fat". Meanwhile, just down the hallway, the lunchroom is totally filled with all the Chinese staff and students, who couldn't countenance the idea of eating at their desk. So if I happen to go into the lunchroom when this is happening I am always greeted with "Ni Hao" and lots of laughter at my attempts to reciprocate in Mandarin. The Chinese find food REALLY important, even to the extent of including many items under the label of "food" that most other cultures baulk at. Amanda Bennett was pretty close to the mark when she said, "Cantonese will eat anything in the sky but airplanes, anything in the sea but submarines, and anything with four legs but the table." And I don't think it's just the Cantonese. On my first visit to China (Yunnan), I was fed fried grasshoppers, "smashed" tortoise, bee larvae, deep-fried bamboo maggots (tasted like McDonalds French fries – so pretty good) and a bright orange fungus that only grows parasitically in caterpillars.

In 1994 I spent a 10-week period doing a small project at an agricultural research institute in Montpellier in the South of France. The first day I was very pleasantly surprised when the entire lab downed tools (well OK, pipettes) on the dot of 12 noon and took me off to the staff restaurant. I was given a card for staff and then we filed along through the food area, piled high with such delights as salade Niçoise, assiette Anglaise, blanquette de veau, tagine with couscous, and an array of mousses, crème caramels, etc., and a glass of wine (or small bottle), all for about €3. We sat around a civilised table, and ate and made conversation, and then some of the younger lab members went outside to play boule (but mostly to have a cigarette), and then we made it back to the lab by 1.00 or so. I thanked them for taking me to such a good place for lunch – this is exactly what we do when we have a visitor to welcome them for the first day. Next day we did exactly the same thing, then the next and so on. *C'est normal!* The French could not abide the concept of sitting at a computer eating a sandwich, and definitely not heading out to the McDonalds drive-thru to get a quarter pounder (Royale with Cheese as John Travolta almost accurately told us in *Pulp Fiction*). Food is French culture, or maybe French culture is food. Or perhaps more eloquently put by Charles Pierre Monselet: "*Food* is a central activity of mankind and one of the single most significant trademarks of a *culture.*" The French are not only immensely proud of this cultural food *chauvinism*, they also utilise it as a means of setting themselves apart. And there is nothing the French like to do more than cast aspersions on British and American food. Francois Tanty, who wrote a cookery book on French cuisine in 1895, adapted to American tastes, wrote in his foreword:

> *I cannot protest enough against the custom so general in the United States to give to the table only the necessary time and to eat like a locomotive taking water, by doing which you expose yourself to the various stomach diseases which make so rapidly the fortune of the doctors and druggists.*

And Rudyard Kipling, although himself British, seemed to be of a similar opinion:

> *The American does not drink at meals as a sensible man should. Indeed, he has no meals. He stuffs for ten minutes thrice a day.*

However, given that a data-free opinion or two from a plant geneticist may not be informative, I talked to some experts on the subject, including a food scientist and a renowned award-winning chef and a social scientist.

Heather Smyth is a fellow UQ academic and food scientist, although perhaps I'm being unfair lumping her in with that group. In my own cereal quality improvement research I get to work with many food scientists – great scientists, don't get me wrong, but they either get off on mathematical equations for modelling starch biosynthesis or incessantly rabbit on about the importance of eating fruit and vegetables, pulses, wholegrains – you know – all those things we know about and choose to studiously ignore when faced with a choice between a stick of celery or a Sachertorte. Those food scientists who are all about guilt – the fun police of nutrition.

In that company, Heather is a breath of fresh air. On her webpage she is described as a flavour chemist and sensory scientist. I've got to admit I'm more than a tad envious of her job, because among other things, she faces the challenge of trying to gain an understanding of consumer enjoyment of coffee, beer and wine (three of the major food groups). Another of her challenges is understanding the chemistry and influence of *terroir* on locally grown produce. I'm not talking about the difference between a stick of spaghetti from Ames, Iowa and one from Parma, or Marmite from Christchurch *versus* from Burton-upon-Trent. No, I mean comparisons such as a Shiraz from McLaren Vale *versus* a Syrah from Crozes Hermitage. Or a lobster from the cool Tasmanian waters *versus* another from the equally cool waters off the coast of Maine. Really? How do you get a gig like that? Finally, somebody undertaking important research on the major food groups – and yes, she also does chocolate (fourth major food group covered!).

Heather did her PhD in chemistry at Adelaide University. Sounds pretty boring, huh? Well not when you hear it was flavour chemistry, and her whole thesis research was to determine the chemistry behind what made Chardonnay and Riesling taste different! The rest of us students were just trying to work out what grape varieties went into our boxes of wine (delightfully known as "goon" in Australia). No that's not true, most students just wanted to know that it was wine, and how many plastic cups they could drink of it before passing out under a bush, or getting up the courage to go talk to that cute guy across the room who was in their geology tutorial group. Shortly after finishing her PhD, Heather moved to Queensland and embarked on trying to find some new challenges, given that the wine grown in Queensland is... how do I put it? Their website says, "it's one of the State's best kept secrets", so let's say it's embryonic but improving. So Heather was looking for a new challenge and it came in the form of seafood.

One weekend she saw a TV story about Lester Murray, from the Eyre Peninsula in South Australia. Lester is a Coffin Bay oyster producer who had been on an international study tour to help him develop good marketing strategies for the region's produce, and he had concluded that the future was to market the *terroir*, but he wasn't sure how to do that effectively. And here's where Heather comes in. She phoned him the very next day to offer her assistance. In Heather's words, the big hurdle was to develop a manner in which to describe the produce accurately and make it sound unique – instead of the general "kinda salty, like an oyster", something more descriptive that nailed the distinct *terroir* (is it *terroir* when it's from the ocean?). The outcome of the collaboration is a brilliant marketing strategy under the banner "Seafood Frontier" (Google it, says Heather). They developed and used a seafood flavour wheel to help producers and consumers to describe the particular seafood they were tasting. And according to Heather, there was a depth and complexity of flavour in the 50 different species of fish and shellfish from the Eyre Peninsula that was accentuated by a unique "greenish fresh herbal flavour" and a tenderness that was unattainable in other regions. Now some of you are reading this and planning your next holiday. Others are thinking "ugghhh – oysters!!".

Heather also gave me some insight into the role human genetic variation can play in the foods we like (and dislike). Some of us are apparently "super-tasters" – we can pick up certain chemical signatures at very low concentrations. We also differ in the enzyme composition and concentration in our saliva, and in our response to food textures. These can have a huge effect on the foods we like and dislike. And with this genetic variation comes population differences, one extreme example being the high proportion of super-tasters among Japanese females. And being a super-taster is not a good thing. When you are a super-taster, you have a higher density of spongiform papillae on your tongue. Typically super-tasters don't like to try new foods, and usually develop a dislike for going out to dinner. Apparently, you can do a blue dye test to assay the density of spongiform papillae on your tongue, but as dining out is one of my greatest pleasures, I don't think I need to do the test.

Another of my learnings from Heather is the probable reason why tofu is not high up in the list of my favourite foods. I've never understood why some people absolutely love it. Again, it's genetics at play. Human genetics and plant genetics in this case. What that means is we have genetic variation for our abilities to taste certain chemical components of the food, and that among and within plant species, there is genetic variation for the chemicals that the food contains. Two of the flavour components that press our like/dislike buttons with soy products are isoflavones and n-hexanal. Now for those readers who think this is getting a bit too scientific, fear not.

For those readers who are thinking, "Wait a minute, I only eat organic, chemical-free food", I hope you are still with me for Chapter 7. But as a brief preview, consider these facts:

- The only thing that is chemical free is a vacuum – that hypothetical vacuum wherein high school physics experiments take place.
- Water is a chemical.
- When you breathe in, it's because you want a particular chemical called oxygen. Actually, it's not a want, it's a need.
- When you breathe out, it's because you want to reduce the concentration of another chemical, carbon dioxide, in your lungs.
- You are made of chemicals, mostly compounds containing the elements hydrogen, carbon, oxygen and nitrogen. But you are also calcium, phosphorus, sulphur, copper, iron, sodium, chlorine, magnesium, zinc, potassium, gold and lots of other elements too.
- If you can smell the coffee from "over here", it's because the coffee "over there" has released chemicals and they have entered your nose.
- If you can taste the sweetness in your mouth, it's because a chemical (sucrose, fructose, aspartame, stevia, whatever that stuff you get in the café that says "sweetener" and spoils the

taste of whatever you add it to) has stimulated a taste bud and sent a "sweet" signal to your brain.

- The active part of the taste bud is a chemoreceptor. A chemoreceptor is a cell made of chemicals that detect other chemicals, and connected to a nerve cell.
- The "sweet" signal to your brain was predominantly electricity. This also involves a bunch of chemicals, with charged ions moving across cell membranes.
- If for some reason you are shocked about being full of chemicals, with millions of chemical reactions going on inside your body, perhaps you can stop reading now and have a lie down. Or get a cup of tea, the chemicals in it will calm you. Or maybe they'll stimulate you. Depends on you. Depends on the tea. Or if it's all still too much, get yourself a "detox" (which will involve adding different chemicals to your body).

But back to those soy chemicals. Turns out there is human genetic variation (like there is for almost everything) and people of European origin are much more likely to taste the n-hexanal and isoflavones in soy products like milk and tofu, whereas among Asian populations, many people cannot detect these flavour compounds, or require higher concentrations to detect them. As soy products age, there is a breakdown of the main oils they contain into n-hexanal, which to many Western palates has an unpleasant "beany" flavour. A group of enzymes, the lipoxygenases, are responsible for this breakdown. Some soybean types carry a natural mutation (in the gene SBL-2) which means they do not produce the lipoxygenase that makes n-hexanal. A group at Iowa State University[1] did sensory testing of soy milk and tofu in a group of graduate students, subdivided into American, Chinese and Japanese. The Japanese students could not detect any appreciable difference between the normal and mutant soy foods. The American students found that the raw beany flavour and aroma was lower in the mutant soy food. The Chinese students said the raw beany flavour and aroma was higher in the mutant soy food.

And as for those super-tasters, many of them don't like soy products because of the isoflavone content. And many also have problems with the bitterness of coffee, saltiness of olives, flavonoids in citrus, polyphenols in green tea and red wine, and the glucosinolates in brassicas like Brussels sprouts and broccoli.[2]

One of Heather's latest assignments was for the local Performing Arts Centre in Brisbane, where she was commissioned to match foods with emotions – the basic premise that the restaurant could serve food in harmony with emotional intentions of the performance, be it a tragic opera, a Shakespearean comedy or modern musical regarding a small girl with really bad parents who is blessed with magical powers. Oh, and did I mention her current research on craft beer, and coffee – what a day job.

Philip Johnson made his name as the head chef and owner of e'cco bistro, one of the iconic restaurants of Australia, which won the Gourmet Traveller Restaurant of the Year Award in 1997. Johnson opened his bistro in 1995 in an old tea warehouse. It has a light airy feel, and an open kitchen where you can see much of the culinary activity, which back 20 years ago seemed something really innovative. Originally a New Zealander, he grew up just outside Christchurch, learning cooking from his mother and grandmother. Coming from three generations of motor garage owners young Philip wanted to be an aircraft mechanic. When he was 16 his application to get an apprenticeship with Air New Zealand was unsuccessful, so his mother walked him into a hotel and on the very same day signed him up as an apprentice chef. The hotel was owned by a group call Ballins, a soft drink company (later taken over by Coca-Cola) who had 15 hotels around New Zealand. "I suppose I was very fortunate as they moved me around so I got a wealth of experiences in different places." By 1983, he found himself chef at Ménage a Trois in London, and on his days off, he worked as a pastry chef at the Ritz for free. Ménage a Trois was somewhat ahead of its time, as AA Gill wrote in *The Sunday Times* in 1997:

His restaurant Ménage a Trois was truly inspirational. Serving only starters and puddings, it became risible only through imitation and the worst excesses of nouvelle cuisine. Thompson is the only British chef to have been made a member of the French Academie Culinaire, and this at a time when British food was still in the dark ages. His descent into celebrity is either a cautionary tale or a parable, depending on where your values are.

I was fortunate enough to talk with Philip while he and his wife, Mary, were frantically remodelling a Spanish tapas bar in Brisbane's Fortitude Valley to a 1950s French restaurant called "Madame Rouge". A new venture and very different from the open airy style at e'cco, Madame Rouge is dark wood, low lighting with heavy red velvet drapes and Johnson's motto at e'cco is, "We do as little as possible with the best ingredients available". Now you may have gathered that I'm a tad partial to seafood, and one of the great selling points of e'cco is the wonderful ways in which they dish up beautiful seafood. Which is why I was incredulous when I asked Philip (my interview technique has not developed much further from that 10-year-old boy) did he have any favourite foods. So he called across the room:

"What is my favourite food, Mary?"

"Anything that's edible."

"Not quite true", he retorted, "and I spose I do have a fascination with desserts. Plus I have always been allergic to seafood, but I have recently grown out of it".

Anyone who has been to e'cco and tasted the wonderful things that Philip does with scallops or Moreton Bay bugs (*Thenus orientalis* – which is apparently a slipper lobster; Figure 1.2) would assume that he was an aficionado. "No, I enjoy cooking

Figure 1.2 *Thenus orientalis* (Morton Bay bugs).
Photo by arhendrix (© Shutterstock).

with seafood, but have to get others to do the tasting. I was once at a wedding and had to give a speech. Before the speech, I popped what I thought was some sort of pork or chicken ball into my mouth. Turns out it was fish and my whole tongue and inside of my mouth swelled up. I could hardly give the speech and slurred so much that people naturally thought I'd had one too many."

Food allergies are real, and they are actually a response of the human immune system to a food. The US Centre for Disease Control estimates that about 4% of adult Americans have a food allergy, and it is recognised that the "Big 8" foods contribute to most allergic responses. In children the most common food allergies are milk, eggs and peanuts, whereas in adults it is most commonly pollen, tree nuts and fish and shellfish. That is interesting enough in itself, because there is strong evidence to show that the majority of children overcome their childhood food allergies, some to then have a happy allergy-free life, while others develop new allergies in their adulthood.

There are also major differences between countries, based on ethnic origins and the cultural aspects of diets. A recent study showed that childhood allergy was most frequent in Australia and Finland. In Australia, far and away the major childhood allergy is to peanuts, with an estimated 3% of children exhibiting this allergy, whereas in Asia the allergy is less than 0.5%. In total contrast, Singaporean children have high-frequency allergies to shellfish (>5%) and eggs (1.8%). And these allergies are serious medical issues, with the most extreme reaction, anaphylaxis, being life-threatening without rapid treatment.

So if you are a child going to school in Australia or Singapore, it's highly likely that at least one of your classmates has an allergy to peanuts or shellfish, so it may be banned from your school. But hang on, we didn't have allergies when we were kids did we? Well at least I don't recall them being in existence. And it is definitely true that they seem to be an increasing phenomenon, with the prevalence of allergy in the USA increasing by 50% from 1997 to 2011. In that same short time period the frequency of peanut allergy has tripled.

However, one of the outcomes of this is that we have become more aware of food allergies, and many among us perceive we have a food allergy when we don't. In the USA, almost 1 in

5 adults perceive they are allergic to wheat, or at least some of the major proteins in wheat known collectively as gluten. We will revisit gluten in later chapters, but it is a simple fact that only those people with coeliac disease have an allergy to gluten. It is generally not as extreme as many other allergies and tends to be a chronic condition that can have major long-term adverse health effects, even increasing the likelihood of certain types of cancers. However, in most populations less than 0.5% of the population have been clinically diagnosed with coeliac disease, although as many as 1 in 4 adults believe they are gluten-intolerant. And the restaurant and café owners have most certainly catered for it, usually by labelling certain foods as "gluten-free" or, in some cases, banning the use of gluten-containing cereals (wheat, barley, rye) from the menu.

Seems everyone has stories about food, and now that you've endured some of mine and others, let's move on. Agriculture has become increasingly complex and efficient over the past few years, and increasingly global. We seem to be getting to a world where we can get any food at any time. When I was growing up, mangoes and cherries were the Christmas fruits (remember, it's summer down under), but now I can buy these at almost any time of the year. And as for asparagus – that used to be something that came in cans. Now it is in abundant supply year-round, and you have to look at the little tag to see whether it is from Australia, Thailand, Mexico or Peru. But none of this has anything to do with GM food. That's all about management, plant breeding, and post-harvest treatments. In the next chapter, we will discuss modern plant breeding and how GM has played an important role since the 1990s. I'll introduce you to some of the pioneers of the techniques used and the excitement that came in the first decade of our ability to genetically engineer plants.

REFERENCES

1. A. V. Torres-Penaranda, Effect of the use of lipoxygenase-free soybean line on the sensory attributes of soymilk and tofu, *PhD Thesis*, 1999, available at: https://lib.dr.iastate.edu/rtd/12177.
2. A. Drewnowski, S. Ahlstrom Henderson and A. Barratt-Fornell, Genetic Taste Markers and Food Preferences, *Drug Metab. Dispos.*, 2001, **29**(4), 535–538.

CHAPTER 2

A Kind of Magic

"We do not need magic to transform our world. We carry all of the power we need inside ourselves already."

J. K. Rowling[1]

Plant breeders are some of the unsung heroes of modern civilisation. Well, I would say that wouldn't I? It's not just that over half my former graduate students are now card-carrying plant breeders, it's because it's true. I don't know of a single other profession more responsible for the ever-diminishing proportion of humankind who are suffering from malnutrition. Well, it goes without saying, but I'll say it anyway, farmers played a major role too.

Since the 1940s, crop yields have improved many times over. Writing in *Science* in 1983, Nobel Laureate Norman Borlaug (the hero of all plant breeders) attributed approximately 50% of the increase to improved genetics, while the other half is down to better management. More educated and informed farmers were key to making this happen. The tools they were given included nitrogen fertilisers courtesy of the Haber–Bosch process and further understanding of the mineral requirements for crop growth and productivity. Farm machinery not only made things faster for planting and harvesting the crop, but also meant that row spacing was not dictated by the width of the horse's

Good Enough to Eat? Next Generation GM Crops
By Ian D. Godwin
© Ian D. Godwin 2019
Published by the Royal Society of Chemistry, www.rsc.org

muscular buttocks. Narrower row spacing meant more plants per hectare, which translates into more yield.

However, the fact that the application of science has enabled plant breeders to encapsulate such a powerful force for good into a tiny little biological propagule, the seed, is one of the wonders of 20th century scientific advances. Sure, in that century we learned how to fly, even as far as the moon, discovered antibiotics, developed ways to see and hear people from the other side of the world, and invented sliced bread. But I hope you will agree that the little action-packed piece of biology, chemistry and physics that scientists could deliver in a single seed remains one of our greatest achievements.

While it is true that the scientific observations and some very prescient theorizing by Gregor Mendel, Charles Darwin and Alfred Wallace underpin the science of genetics, it is also true that harnessing the understanding of genetics has been key to driving agricultural productivity. Interestingly, both Mendel and Darwin were active participants in the local horticultural societies, and part of their interest in genetics arose from attempting to understand the nature of genetic variation and the inheritance of traits.

Darwin first mentioned some of the issues of plant breeding in his 1868 book *The Variation of Animals and Plants Under Domestication*. One of the plant breeding challenges he found most intriguing was the causal organism of potato late blight (Figure 2.1), *Phytophthora infestans*, which in the 1840s led to the great Irish potato famine, in which many died from starvation. It also led to the Irish diaspora and my relatives, the Kellys and Flanagans, coming to Australia. OK, the Kelly was transported as a convict, but it's my prerogative to maintain he only stole the sheep because he was starving. Darwin actually saw the disease at first hand in 1845, when some of his own home-grown potatoes succumbed to the disease in Kent. He wrote in a letter that it was a "painfully interesting subject",[2] and was soon conjecturing that there must be some genetic variation for the disease, perhaps from some of the varieties he collected himself while voyaging on *The Beagle*. It was not the case, but he was a firm believer that there must be variation for response to the disease in some "wild stock", a concept that still underpins much of modern plant breeding – and the very reason why we are

Figure 2.1 *Phytophthora infestans* (potato late blight), affected plant leaves showing discolouration (photo by Andre Drenth).

keeping hundreds of thousands of crop landraces and wild relatives under a frozen mountain in Svalbard, Norway (Figure 2.2a and b).

Darwin actually pleaded for government funding for a potato breeding programme to overcome the ravages of the disease. In the 1870s he corresponded with James Torbitt, who was among other things a potato breeder, making crosses and growing potatoes from real seed. He actually grew some of Torbitt's varieties in Kent, and while some were disease-resistant, their yields were less than impressive. After failing to convince the government, Darwin eventually funded the breeding programme himself, and talked a number of family members and friends into contributing. In the long run, whenever a new variety was released with resistance, this sneaky pathogen utilised mutation and/or genetic recombination of its own to overcome the resistance. Even today, more fungicides (in excess of $1 billion per annum – that's right, I said billion) are applied to potatoes for the control of late blight than in any other crop-pathogen system.

Once into the 20th century, and Mendel's laws of genetics were "rediscovered" by a number of European groups, plant breeders started to get scientific. They now understood the basics of genetics, or heredity as it was previously known, and the ability to more accurately predict the outcomes of particular crosses became a tangible reality. Tangible, but often unattainable, not

Figure 2.2 (a) Entrance to the Svalbard Global Seed Vault and (b) worker and storage within it. The vault houses 840 000 samples from around the globe.
Image credit, Crop Trust.

least of which because there were many aspects of genetics they did not fully understand. We know a lot more about these now in the age of molecular biology, but suffice to say there are still things we don't have a good handle on. We do understand that when you inbreed a species, you can develop inbreds or pure lines. Now in general, inbreeding is not particularly advantageous in species where outcrossing is mandatory (for example, when you have separate male and female organisms, like humans). However, many plant species have male and female flowers on the same plant, and in many cases, these are even in the same flower. So, nature being what nature is, the most efficient manner in which pollen can effect pollination of a female flower part is on the same flower, or failing that, on the same plant! A plant breeder calls this "selfing", and it is pretty easy to accomplish by putting a physical barrier such as a paper bag around the flowers. This has the dual advantage of preventing pollen from leaving and preventing other pollen from intruding, or in the case of plants that prefer threesomes, it keeps the birds and the bees away. Oh, and the flies and the moths and the beetles and the possums and all those other pollinating organisms. The male organs on the flower produce pollen, and the female parts have eggs or ovules, usually deep inside the flower.

Now, for a potato, you can actually plant the potato tuber and get the same genotype come up. So if you plant an oval-shaped purple-skinned large potato like Royal Blue, you are very unlikely to harvest white-skinned fingerlings like Kipfler 4 months later. That's because sexual reproduction was not involved and hence the potato tuber is a "clone" of the original parent. However, if you plant a bunch of purple maize seeds, you may end up with some purple, red, yellow and white progenies unless the parent line was an inbred. This is known as genetic segregation. Segregation occurs because your purple-seeded maize is the result of sex, and unless you put a bag over the head, or you know that the seed source is inbred, so that all pollen carry the gene for purple seed, you will get segregation. Now farmers had a pretty good handle on this, so they made selections over generations for a variety that was true to type.

This was one of the transforming effects of understanding genetics on agriculture and food production. However, because of the vagaries of plant reproduction, in many species this was hard to get completely right. There was always an element of outcrossing, and as a result, there were always some genes that were segregating. These "almost but not quite inbred lines" are called open-pollinated varieties or landraces. So there is always a small amount of variation in seed colour, or how long it took the plant to mature and set seed, or plant height or disease resistance. So as farmers kept a small proportion of this year's harvest for next year's planting, the open-pollinated varieties always threw up a bit of variation. Of course, from this novel and generally unwanted variation came some fantastic newly improved varieties. If a leaf disease comes through your crop and totally defoliates it, there is always the possibility that there is one green, resistant plant left in the field, and that's how many of the first disease-resistant varieties were selected.

During the 20th century, the two truly transforming genetic technologies for agriculture were: (i) harnessing hybrid vigour and (ii) the Green Revolution. They both led to outstanding increases in food productivity, although much of this has been seen in the cereal crops, leading to wheat, rice and maize being the source of half our human calorific intake.

One of the great challenges of genetics is the understanding of hybrid vigour, or heterosis. Hybrid vigour is used in a number of crops because we know that if you make a combination between two elite inbred lines of crops like the major cereals, or brassicas, or tomatoes, in most cases the ensuing hybrid will out-yield the parents by as much as 20%. This was first realised and understood early in the 20th century, but was not effectively harnessed until the 1930s. Why? Because to create these hybrids, you had to physically cross the selected male and female plants. This was easiest for maize, because the male flowers are at the top of the plant (the tassel) and the female ovules are inside the seed deep inside the cob halfway up the plant, and the female receptive part is called Rapunzel. No that's not true, but she does let down her golden hair, and these are the receptive female parts known collectively as the silk. For a cross, you make a plant a female by de-tasselling (also called emasculation, which somehow sounds so much more threatening) before pollen is formed. So by cutting off the tassel from the male plant in the morning and shaking the pollen from it onto a silk of the female, then sealing it with a bag to prevent any other pollen from getting in on the act, a cross is made. The seed from that cross is known as F1 hybrid seed and if the parents have good "combining ability" the hybrids will yield significantly more than either parent. So farmers will pay more for hybrid seed, to get a 10-tonne crop rather than a 7-tonne crop from an inbred. But this is expensive and laborious seed to produce, and in most other species, just not economically viable. The reason is that it is rarely as easy as de-tasselling to emasculate the female.

Enter Henry A. Wallace, an Iowan with an interest in plants and agriculture, among other things. After attending Iowa State College, he made and sold seeds of Cooper Cross, the first hybrid seed to make it to agriculture, in 1923. He established the Hi-Bred Corn Company in 1926, being renamed Pioneer Hi-Bred in 1935, and by 1939, 90% of all corn grown in Iowa was F1 hybrid seed. However, he was called to Washington as Secretary of Agriculture from 1933 to 1940, then served as Vice-President under Franklin D. Roosevelt from 1941 to 1945.

From 1860 until 1930, average maize yields in the USA were a relatively constant 20–28 bushels per acre. Since the 1930s, maize yields have increased every year from around 30 bushels per acre to 60 by 1960, 100 by 1980 to 140 by 2000. It needs to be said that there were two other non-genetic technologies that also played a role in this success story. The Haber–Bosch process, which allowed nitrogen from the atmosphere to be fixed as ammonia, meant that a cheap source of plant-available nitrogen was now available to farmers. Fertiliser derived from the process became widely available to farmers by 1940. The other non-genetic technology was the replacement of the horse with the tractor. And not because things got faster, they just got a whole lot more flexible. Farmers growing maize in the early 1900s planted maize with 40 inch (100 cm) spacing between rows. Any narrower and the horse just did not fit between the rows for any sort of cultivation or harvesting. This allowed farmers to reduce row spacing and plant more plants per hectare. And with better technology they could actually change the row spacing depending on soil type, rainfall, the variety of crop, fertility, *etc.* And again, it was around the 1940s that tractors started to become widely adopted in Western agriculture (Figure 2.3).

Wallace took a very liberal view of the world, and while Secretary for Agriculture he reorganised US agriculture as part of the New Deal. It is believed that Roosevelt wanted him as Vice-President because he saw him as a worthy successor. While Vice-President, he saw the end of World War Two as allowing for substantial change to the world order, being widely seen as too pro-Soviet, too idealistic and against colonial empires. Even

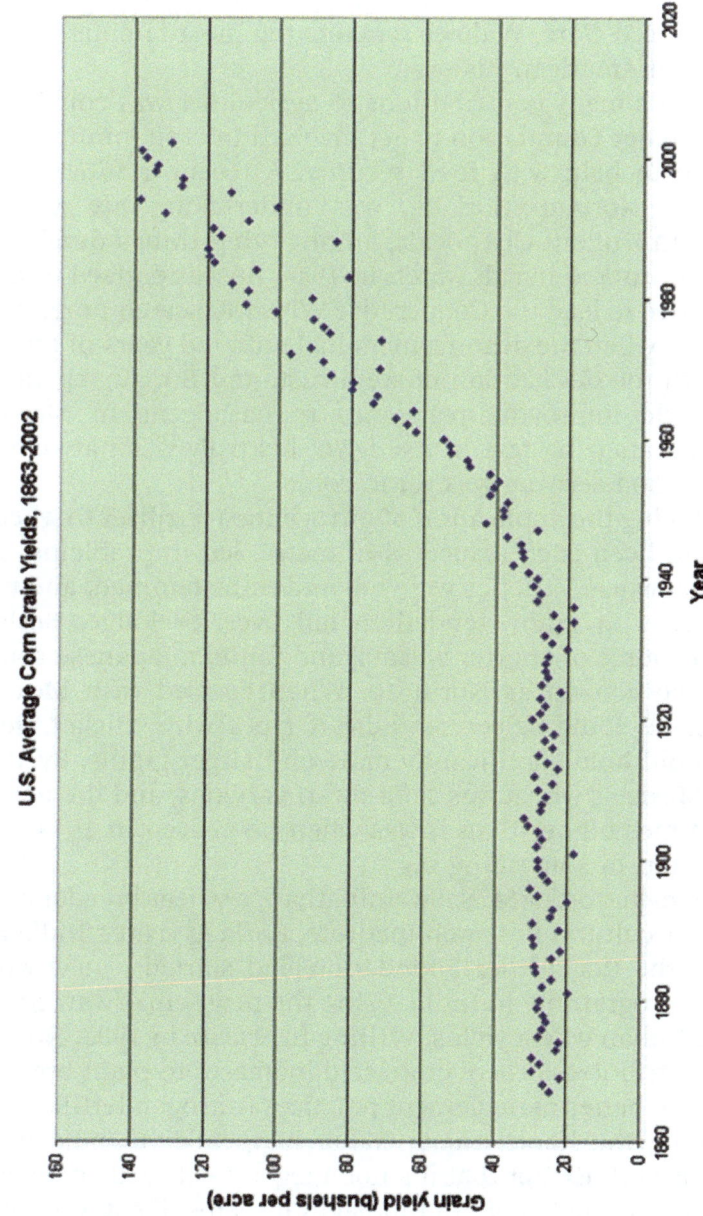

Figure 2.3 Average maize (corn) yields in the USA from 1860 to 2002. Yields lower than the trend line are usually the result of drought and/or heat stress.

among conservative Democrats this was all a little too radical, and in 1945 he was replaced by Harry S. Truman as Vice-President. As Alex Ross wrote in *The New Yorker*, "With the exception of Al Gore, Wallace remains the most famous almost-President in American history."

Among his many contributions to agriculture was convincing the Rockefeller Foundation to get involved in crop improvement in Mexico to help with food security. A recent graduate from Minnesota, Norman Borlaug, was undertaking his wartime service with DuPont Chemicals, among other things developing a glue that worked in salt water. In 1944, he was enticed to head off to Mexico to lead the Cooperative Wheat Research Program. It was a time when Mexican farmers had suffered years of terrible yields with the devastation of stem rust, and Borlaug spent the next decade improving resistance to pathogens in Mexican wheats, an amazing feat in itself. Yet his truly visionary contribution to food security was yet to come.

Introducing the application of nitrogenous fertiliser to wheats, which had been such a success in maize, led to problems. The traditional wheat varieties were tall and thin-stemmed, and with the increase in grain, tended to fall over, or lodge. Borlaug sought a source of shorter wheats, and found a Japanese source in a variety known as Norin 10. When crossed with Mexican varieties, he could select semi-dwarf types with thicker stems which could hold significantly more and larger grains. By 1963, 95% of Mexico's wheat was semi-dwarf varieties, and the harvest was six times higher than it was when he arrived in 1944. This was the start of something big.

At the invitation of M. S. Swaminathan, a wheat breeder at the Indian Agricultural Research Institute, Borlaug visited India and by 1963, the Rockefeller Foundation had started a new wheat breeding programme in India, using the new semi-dwarf trait to improve Indian wheat yields. Writing in *Science* in 1983, Norman Borlaug attributed 50% of crop yield increases to plant breeding and 50% to better management practices (nitrogen fertiliser and overcoming the horse's arse). What happened in India is the stuff of legend. Except that it's not a legend – it really happened and even with the benefit of hindsight it seems like it was a kind of magic. After five short years in India, the magic was evident. In 1966/67 wheat production in India was just over 11 million tonnes. Within 2 years, in what became known as the Green Revolution, wheat production had doubled, and by 1980/81 the harvest was a staggering 36.5 million tonnes, which provided the carbohydrate needs for an extra 186 million people. What is truly amazing is that for most modern cultivars of wheat, the average turnaround for a new variety to be developed is in the order of a decade, yet the Green Revolution managed to double wheat productivity in under 7 years.

Norman Borlaug was awarded the Nobel Peace Prize in 1970 by the Norwegian Nobel Committee. In his acceptance speech he stated:

"The Green Revolution has won a temporary success in man's war against hunger and deprivation; it has given man a breathing space. If fully implemented, the revolution can provide sufficient food for sustenance during the next three decades. But the frightening power of human reproduction must also be curbed; otherwise the success of the Green Revolution will be ephemeral only."

In 1986 Borlaug instigated the prestigious World Food Prize and the first recipient was his Indian colleague, M. S. Swaminathan.

When I went to the University of Queensland in the early 1980s to study agricultural science, I was pretty certain I wanted to be a plant breeder. Of course that wasn't always the case. My first true aspiration was to be an Olympic high jumper (too short, among other issues), then a flying doctor, until I found out that the pilot wasn't the doctor, and *vice versa*. Then in the midst of a hitherto boring subject called biology, we did a semester of genetics, and suddenly biology became mechanistic and something that you could apply and "do stuff with". And then we had a visiting scientist. A tomato breeder came in one day to give us a talk on plant breeding and that was it – that's what I wanted to be.

So, in the early 1980s I was following all of this avidly. No, that's not true. I was oblivious to all of this stuff about the Green Revolution. I was more interested in following girls, my football team and politics. But the more genetics I learned, the more I got to understand that not only was plant genetics pretty interesting, it was also a major transformative force for humanity.

And speaking of transformation, that was the next revolution in agriculture and food production. Today we talk about transforming industries, transforming our political system, transforming society with disruptive technologies, but to plant scientists in the early 1980s, plant genetic transformation was the future, and it was exciting.

The promise of genetic transformation was that we could take a gene from anywhere – even a gene outside the plant kingdom (such as from an animal, a bacterium, a virus), and get that gene permanently incorporated into a plant's genetic make-up and change that plant. Why would we want to do that? Perhaps there was a disease that affected our favourite species – for argument's sake let's say a potato. And having looked at all the available potato genetic variability, there was not a single source of resistance to be found. Potato belongs to the *Solanum* genome, which is not only where many potato relatives live, but included things like *Solanum belladonna* (deadly nightshade) and perhaps more palatable, *Solanum lycopersici*. This species used to be known as *Lycopersicum esculentum*, which means the edible wolf peach (Figures 2.4 and 2.5). You've all eaten it, many on an almost daily basis. It's what we call the tomato, or tomayto if you are North American.

So what if the tomato had resistance to a plant disease that the potato did not? Could we use this gene in the potato? Well

Figure 2.4   *Solanum lycopersici*, tomato varieties.

Figure 2.5 *Solanum lycopersici*, tomatoes unripe on the vine.

unfortunately, potato and tomato are what we call sexually incompatible, so we can't transfer genes from one to the other. But what if we could somehow get a copy of the tomato disease resistance gene and transfer that single gene to the potato *via* genetic transformation? I even remember asking my plant breeding Professor, the late Don Byth, this question when I was an undergraduate. He said "well it may happen in your lifetime sonny, but not likely in mine". That was 1982.

The term "genetic transformation" was first demonstrated in 1928 by British microbiologist Frederick Griffith. He heat killed a virulent strain of *Salmonella pneumoniae*, then exposed a non-virulent strain to the remnants of the dead strain, and the non-virulent strain became virulent. He hypothesised there was some form of "transforming principle", the nature of which was not clear. Tragically, Griffith was killed in the London Blitz in 1941. Famously, Oswald Avery and colleagues at Rockefeller University, New York, were able to demonstrate in 1944 that the transforming principle was indeed DNA. We now take it for granted that bacteria can take up circular pieces of DNA known as plasmids, and if these plasmids confer a "selective advantage" *a la* Darwin, they will be maintained by the bacterial population. One such typical advantage is a resistance to antibiotics. Hence when we apply an antibiotic to a bacterial population, we are killing off those cells that do not contain a plasmid with a bacterial gene in them. If all cells don't have the plasmid, the entire population will be wiped out. However, all we need is one cell with a resistance gene on its plasmid and that cell has a selective advantage. We have created a situation where the cell with antibiotic resistance can not only survive the toxic chemical, but we have also removed its competition (all those other cells). That single antibiotic cell will go forth and multiply, with quite terrifying speed. And the result of that is that an ever-increasing proportion of the population will carry the resistance gene.

Of course, this all makes total sense to high-school biology students. They can see bacterial cells doing all sorts of things on YouTube videos. But back in the 1940s there was in no way a shortage of sceptics. Let's not forget that this was around the time when infections were serious problems in surgery and in wounds, and we did not have particularly good weapons against them. However, by the 1970s people like Stan Cohen at Stanford had demonstrated that you can get bacteria like *Escherichia coli* to take up plasmids by treating the cells with calcium chloride at a specific concentration. And naturally, this led to speculation that if bacterial cells could do this, why not plant, animal and even human cells (yes, humans are indeed an animal, you are quite correct). And with a burgeoning biotechnology industry in the late 1970s, molecular biologists started to think that it was not only possible, but it rapidly became an international race among biologists. Not as exciting as NASA *vs.* the Soviet Union, but mainly because you couldn't watch it on television.

It's not surprising that the most skilful and knowledgeable molecular biologists of the late 1970s were mostly microbiologists. Yet they were stymied by a few not so subtle differences between microbes and plants. Until this time, all transformations had been done with single bacterial cells. And it was relatively simple to get a single "transformed" cell to go forth and multiply. A microbiologist could inoculate a small number of transformed cells into a 1-litre flask of broth (chemically very simple – like vegemite + sugar), shake this broth overnight at 37 °C, and then the next morning you will find more than a billion of these cells per millilitre of broth. Remember what I said above about terrifying rates of population increase? So how do we get this to work with plants or animals, which are multicellular organisms, with many different cell types? As you can imagine, different groups took different inventive approaches, for many and varied reasons.

While there were many groups around the world all striving to be the first to make a transformed plant, I am going to illustrate the excitement of the times with what was happening in four groups. The Cocking group at Nottingham, the van Montagu/ Schell group at Ghent in Belgium, the new biotech company, Monsanto, in St Louis, and a little maize seed company, DeKalb in their labs in Mystic, Connecticut, where Julia Roberts famously served pizza with that infectious smile. What, oh that was only in a movie? I knew that.

Ed Cocking's group in Nottingham decided that they would take the approach of dividing plant cells into single cells. Not as easy as it sounds ... well it doesn't even sound easy, let's get real. Plant cells are organised and held together in complex structures like roots, leaves and flowers by something that animal and plant cells don't have. Plant Cell Walls. Cell walls are made up of long interwoven fibres of polymers known as cellulose, hemicellulose and lignin. Or what the nutritionists call dietary fibre. It's supposed to be really healthy, which is why we are constantly being told to eat 200 pieces of fruit and 750 vegetables every day! We're all trying, aren't we? Surely there are still plenty of cell walls in wheat beer and potato chips?

So anyway, to break plant cells into single units we can use enzymes, and these yield "protoplasts", which are unicellular. Ed Cocking first demonstrated that this was possible back as early as 1960. So these protoplasts resemble microbial cells, and surely we can coax them into taking up foreign DNA. Well, yes, we can. Going back to Stan Cohen's examples of making competent cells, we can use all sorts of inducements – cold, ionic shock, electricity – and lo and behold, DNA enters the cell. If we are lucky it actually enters the nucleus of the cell, and may get incorporated permanently into a chromosome. So now, here is a single, naked, transformed protoplast and all we need to do is turn it back into a plant.

That's where the real problems begin. If the protoplast can't be induced to make a cell wall and undergo cell division, then that's about it. A one-way ticket to a sad, lonely, unicellular death, often watched by a student who is peering hopefully down a microscope to see their next potential ground-breaking paper shrivel and die before their eyes (it still hurts, believe me). I wish I could stop saying believe me, believe me.

The Cocking group, among others, were able to show that certain plants could get through these traumas, and form cell walls, divide and develop into whole plants. So as a result, it was hypothetically possible that given the right treatment, these protoplasts could uptake foreign DNA and then go on to form a transgenic plant. One plant in particular was really good at regenerating whole plants, and it became the darling of molecular biologists in the late 1970s and through the 1980s. Never mind that it may well have killed more humans on the planet than any other plant or animal species. That plant is tobacco. A native American taken to Europe in the 1600s to be harnessed by humans to start its killing spree, which still continues in the 21st century. But I digress.

Tobacco (*Nicotiana tabacum*) demonstrated that it was remarkably amenable to regenerating whole plants from protoplasts, single cells, leaf pieces. It became the workhorse of plant MoBo labs, and we still use it today in undergraduate student practical classes because "anyone can do it". The Cocking lab had shown that they could regenerate tobacco protoplasts into whole plants, so they were focussed on using similar techniques to the microbiologists in producing transgenic tobacco.

A terribly funny Irish colleague of mine, John Hamill, was doing his PhD in the Cocking lab in the early 1980s. John is now a Professor at Deakin University, but back in the day – he knew how to party. In fact, I had only known him for 24 hours when we met at conference in Elsinore, just north of Copenhagen. We were room-mates. Elsinore is where Shakespeare set Hamlet, and something was indeed rotten in the state of Denmark after the first night of partying. After I'd gone to bed and had been sleeping some time, John appeared at the bottom of my bed with a skull in his hand, ready for a re-enactment. I sat bolt upright and was expecting something about poor Yorick. But what came out of his mouth, in his Belfast accent, was more along the lines of "yergoddenyyfeckkinahsprin". After about ten attempts, I finally woke up enough to realise that he didn't have a skull in his hand, but a glass of water and he wanted aspirin. It was the start of a lifelong friendship.

John was full of stories about his time in the Cocking lab. His project was based on the use of protoplasts, but not as an actual means of plant transformation, but an asymmetric gene transfer *via* protoplast fusion. Protoplast fusion was a pretty hot topic at the time, with many labs wanting to show they could potentially fuse cells of a potato (*Solanum tuberosum*; Figure 2.6) and a tomato (*Solanum lycopersici*) to get a plant which produced potatoes below ground and tomatoes above ground. Meanwhile others in the lab were engaged in the race to produce the first transgenic plants, by introducing naked DNA (plasmids) into protoplasts, and hoping they could regenerate whole plants from the protoplasts.

Now is probably a good time to talk about DNA and genes. If you haven't seen images of the DNA double helix by now, you must have been living under a rock. And if you were living under a rock, the only reason you couldn't see the DNA was because it was microscopic. Because there was DNA under that rock too. And there's no need to tut-tut either – no you really couldn't help yourself, could you? You've just loosened a few cells from inside your mouth and swallowed them, which technically means you have ingested DNA. Well, and animal cells. In fact, you may have just committed cannibalism, albeit on a microscopic cellular scale.

By the late 1970s there was an emerging understanding of how genes worked. And if you really want to know how genes work

Figure 2.6 *Solanum tuberosum*, potato varieties.

ask any geneticist. And then ask another one, and they'll have a slightly different answer. In reality, we understand the basics quite well, but the subtleties are a different thing. Since Watson and Crick first worked out the chemical structure of the DNA helix, after famously interpreting some of Rosalind Franklin's X-ray diffraction pictures on a visit to London, we have known that the two strands of DNA are made up of a four-letter alphabet. This alphabet is A, C, G and T, and stands for the nucleotides adenine, cytosine, guanine and thymine. These nucleotides are joined together in a large strand, and twisted around another large strand (the complementary strand) into a helix, then held together by hydrogen bonds in nucleotide or base-pairs. A always pairs with T and C always with G.

So here goes with the basics of gene structure and gene expression. And we are going to use LEGO® blocks as our building blocks.

So imagine you have a gene from a tomato, made of LEGO blocks, which for illustrative purposes are red. A simple gene will have three blocks, the first one, the middle one, and the last one. When many people think of a gene (often geneticists included), they are really only thinking of the middle block, which is known as the coding region. It is called the coding region because it is actually a template for the complex but really elegant process known as gene expression. The template consists of a code (with that limited four-letter alphabet) made up of three-letter "codons". If you feel lost at this stage, then just skip ahead to the LEGO block pictures (Figure 2.7) because the understanding of this code is neither crucial, nor particularly interesting to most non-scientists. All you need to know is that the three-letter genetic code encodes an amino acid sequence. So for example, ATG TAT GAC encodes three amino acids, methionine–tryptophan–asparagine. So a total coding sequence of 3000 nucleotides can be transcribed in RNA (a single-stranded intermediate) then translated into a linear chain of 1000 amino acids, which is the basis for a protein. Stick ten amino acids together in a chain and you have a peptide. Once the peptide becomes a lot longer it's a polypeptide. Polypeptides usually require some modification to then become a protein. Sometimes they need to fold in a particular way, or join with another protein, or get a metal or phosphate added to them, or head off to a specific place

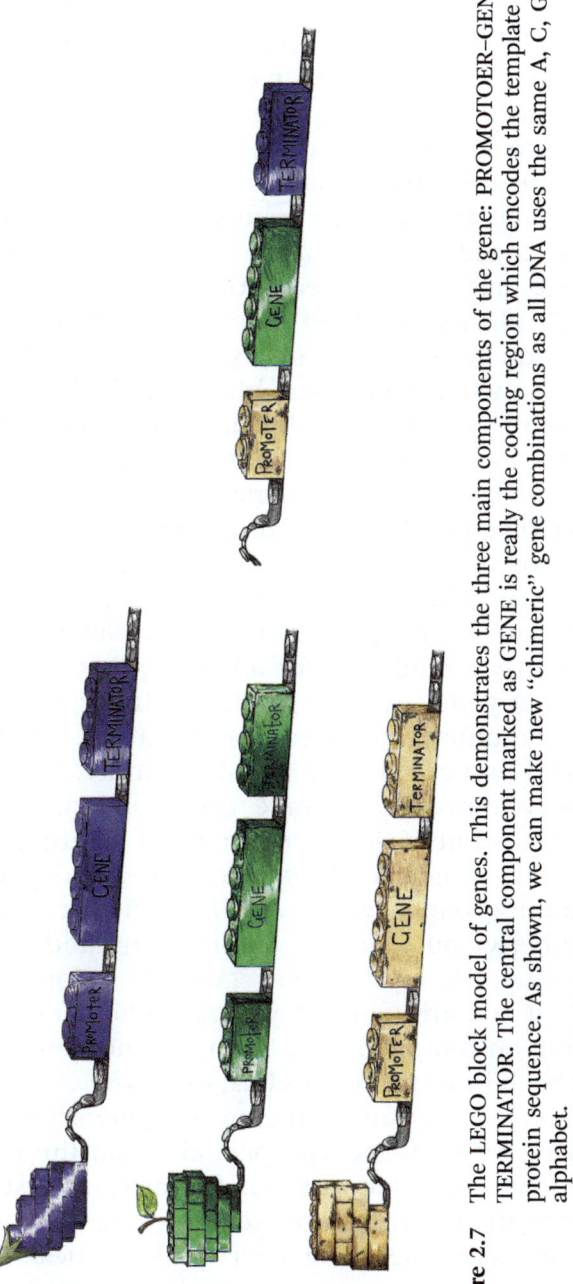

Figure 2.7 The LEGO block model of genes. This demonstrates the three main components of the gene: PROMOTOER–GENE–TERMINATOR. The central component marked as GENE is really the coding region which encodes the template for protein sequence. As shown, we can make new "chimeric" gene combinations as all DNA uses the same A, C, G, T alphabet.

Image credit, Lara-Simone Pretorius.

in the cell to become further processed. And many of them are enzymes, which catalyse chemical reactions in the cell to make new chemicals like other proteins, sugars, fats, and even structural components – things like hair, fingernails, corneas, skin, which are all variants of one protein, keratin.

And then to labour the point, in many plant and animal genomes there are about 25 000–30 000 of these genes and they make up to 100 000 different proteins. And all these proteins have been encoded by one or more coding regions of a gene. So why is a gene made up of three LEGO blocks?

The LEGO blocks on either side of the coding regions are regulatory sequences. We geneticists call them the upstream regulatory sequence (the first block) and the downstream regulatory sequence (the last block), when we are being proper scientists, but in general they are referred to as the promoter and the terminator, based on the general and pretty much inaccurate assumption that the promoter turns on gene expression and the terminator turns off gene expression. Details, details. But what we can say the regulatory sequences do is control gene expression. And that's pretty important. Gene expression is generally tightly modulated, for all sorts of good reasons. It's important that genes get upregulated at the right time, in the right place and under the correct circumstances. Some genes are regulated in a tissue- or organ- specific manner. Some genes are regulated developmentally. Some genes are turned on/off by environmental stimuli. Going back to those keratin genes, if you get the expression right you have hair on the top of your cranium and not coming out of your eyeballs. And then when you go through puberty you wake up one morning and hair appears to be growing where it wasn't growing before. And then, particularly if you are male, as you get older it can sometimes stop growing on top of your cranium, but nothing seems to be holding it back in your nose and ears – ewww!

And as I write this I am changing the gene expression in my liver. Tucking into a wonderful pale ale from the Lord Nelson pub in the Rocks in Sydney (waiting for the sponsorship deals to come pouring in), I have upregulated the expression of a number of genes, mostly my alcohol dehydrogenase genes. These genes are producing an enzyme to convert the alcohol in my bloodstream, and they have been switched into a much higher level of expression after the detection of alcohol in my blood. Well, at least I hope they have been or I'm going to die from this toxin. Don't panic, I've asked Dr Google and he/she says that I would have to drink 13 shots of 40% alcohol to have a 50% chance of death by alcohol poisoning. But more on that later. Lots of genes have changes in expression (up or down) in response to an outside or internal stimulus (temperature, light, chemicals, hormones, radiation, *etc.*). And this is how organisms function. In fact, it has been demonstrated that simply by spraying a plant with water, you can change the expression of thousands of genes.

So back to that tomato gene, represented by the three red LEGO blocks. Now imagine that this gene was a gene that gave resistance to a nematode that made large nodular growths on a tomato root (root knot nematode), which restricted the ability of the tomato roots to take up water and nutrients from the soil. As a result, when the plant becomes sufficiently infected, the plant wilts and dies from desiccation. But that's OK because we have our red LEGO blocks and we can cross the plant with this resistance gene to the plant that is susceptible to root knot nematode and produce new resistant varieties with the red LEGO gene.

Now, imagine the same disease also occurs in potato, but no matter how much we search, we cannot find a source of resistance (in other words a resistance gene) in the potato. And we cannot cross a tomato with a potato *via* conventional plant breeding. And it is pretty frustrating to know we have a resistance gene, just not in the plant where we really need it.

So what if we could take the gene from the tomato and get it into a test tube? Yes, I know all you young clever biologists, nobody uses test tubes anymore. But nobody reads newspapers anymore and we still talk about the press. OK, so we get the gene into a small plastic microfuge tube, and then clone it (copy it) into a plasmid so we can grow billions of copies of the gene overnight in *E. coli*. What if we transform it into a potato? Can we get a potato to be nematode resistant? Yes We Can. But what if the tomato gene is expressed so that the resistant protein is in all the below-ground tissues? Maybe we would like it to be in roots only, and not in all below-ground tissues such as the potato part we eat – the tuber. Maybe that's not so difficult because the tuber is botanically not a root at all – it's a thickened modified stem. So we can also clone a promoter which is root specific from the potato, and replace the tomato promoter. We now have a LEGO block gene with a white promoter from the potato, a red middle coding region from the tomato and a red end region from the tomato. This is what we call a chimeric gene, based on the Greek *Chimaera*, which was a mythical hybrid, and was biologically gifted with a lion's head, a goat's head in the middle of its back and a snake for a tail. So in reality we can swap all these parts of the gene around to make all sorts of chimeric genes. The first transformed insect-resistant plants had a LEGO block gene from three totally different non-plant sources: a coding region from an insecticidal toxin-producing bacterium, *Bacillus thuringiensis*, a promoter from a cauliflower mosaic virus, and a terminator from another bacterium, *Agrobacterium tumefaciens.*

So back to Nottingham in the early 1980s. The Cocking lab was a tissue culture and biochemistry lab and lacked any hard-core MoBo "gene jockeys". So they had plasmids made to order from labs in Cambridge and Norwich delivered by courier on a weekly basis. The plasmids had an antibiotic resistance gene, a gene called neomycin phosphotransferase (*nptII*), isolated from a small mobile piece of DNA (a transposon), which was found on a plasmid in *E. coli*. The neomycin-type antibiotics include kana-mycin and geneticin, both of which are toxic to growing tobacco cells. So following on the plan of what worked with microbes, the aim was to get the kanamycin-resistance gene into the tobacco genome and then culture the cells on a medium containing what would normally be a toxic level of kanamycin. Any cells that grow could be reasonably surmised to contain the kanamycin-resistance gene, and hence be transformed. When Mike Bevan visited from the John Innes Institute in Norwich and gave a seminar regarding transformation systems and mentioned the difference between prokaryotic (single cells like bacteria) and eukaryotic (animals, plants, *etc.*) promoters there was a real-isation in the Cocking lab that maybe they did not have the right plasmids to transform plants with. Protoplasts would eventually be used to make transgenic tobacco plants, but there was another system that had excited many of the microbiologists, and that was harnessing "nature's genetic engineer", a soil-borne bacterium called *Agrobacterium tumefaciens*. *Agrobacterium tumefaciens* had long been recognised as the cause of crown gall disease, a disease that caused plant cells to undergo a neoplastic growth, and form a tumour or crown gall of dis-organised and rapidly dividing cells. An intriguing observation was made, and quite a stunning realisation that even after the bacterium was no longer in the plant, the plant cells continued to grow in an unusual way. Somehow it seemed that the pathogen might be changing the plant genome in some way. The race to genetically engineer or transform a plant resulted in what was pretty much a dead heat, and the recipients of the 2012 World Food Prize, Marc van Montagu, Mary-Dell Chilton and Robb Fraley, all played key roles in this transforming technology.

Marc van Montagu was a microbiologist working at Ghent University in Belgium, and had formed a long-term and fruitful collaboration with Jeff Schell from the Max Planck Institute in Cologne. This man, who was responsible for leading one of the revolutions of agriculture and food production, had very humble beginnings. Quoting from an interview published in 2011 in the *Annual Review of Plant Biology*:

> "The neighbourhood I grew up in was typical for a city relying on a flourishing textile industry: large factories surrounded by a network of dead-end alleys with small working-class houses. Most houses did not have running water; there was a central tap in the street. Some even had common toilets in the middle of the street. Light came from petroleum or gas lamps; few houses had electricity. Heating was done mostly with a coal stove, which also served for the cooking. Since the bedrooms were not heated, there were fascinating ice flowers on the windows during winter."

So one would imagine that he didn't have too many privileges when it came to becoming a scientist, but his family had made a few observations as to the likelihood that a life of manual labour or playing professional football probably weren't likely outcomes. It was a time when very few people actually completed high school.

> "Going on with my education was a big decision, but the family concluded that it was worth trying, because everybody noticed that the only thing I seemed able to do was to read books all day long."

He chose to go on to Ghent University, and study chemistry. After all, Ghent had a long and distinguished history in the chemical sciences, with distinguished former chemists such as August Kekulé (structure of benzene), Adolf von Bayer (the synthesis of aspirin, barbiturates), and Leo Baekeland (the invention of Bakelite) on their faculty. He rapidly changed into a molecular geneticist working on viral particles called phages, which with the growing field of molecular biology could be shown to infect bacteria and use the bacterial cell as a means to reproduce. His first breakthrough was with colleague Jeff Schell, when they were the first to demonstrate that the plant pathogen, *Agrobacterium tumefaciens*, was able to change the way in which plant cells grew by transferring part of their own DNA into the plant. Schell and van Montagu showed that the "transforming factor" in this case was a plasmid, a tumour-inducing Ti plasmid. Bacterial cells could be "cured" of the Ti plasmid and they could no longer infect a plant. You could make the bacterium virulent again by adding the Ti plasmid back into the bacterium. And of course, just like in other bacteria the transforming factor was DNA. However, this was a totally new field. It had already emerged that bacterial cells can exchange DNA in the form of plasmids, but the plant genome was another thing altogether. Unlike bacteria, plants don't have plasmids in their nucleus. Here was the tantalising possibility that maybe, perhaps maybe, this bacterium had evolved the ability to transfer DNA into a plant cell, and even somewhat more fanciful, that the DNA was targeted to the nucleus and through some unknown mechanism, incorporated into a plant chromosome. That seemed far-fetched, but how else could the plant cells still maintain their neoplastic growth in the absence of the pathogen?

And of course, as soon as it became evident that a bacterium could transfer genes, indeed, whole biochemical pathways, into plants, the race was on to harness this natural genetic engineer to transfer something else to a plant. And by something else, I mean something useful, like a trait such as insect resistance or disease resistance. And that's exactly what the van Montagu–Schell group did in Belgium. They had a few competitors too, such as Eugene Nestor's group on the west coast of the USA where Mary-Dell Chilton was working and the newly-formed agbiotech companies, such as Monsanto in St Louis.

It was while undertaking a postdoc in the Nestor lab at University of Washington in Seattle that Mary-Dell Chilton actually demonstrated in 1977 that a fragment of *Agrobacterium* DNA was present in the genome of cells that had the crown gall disease. This fragment of DNA became known as the T-DNA, the transfer DNA. The T-DNA is a small fragment of the Ti plasmid and this explains why in *Agrobacterium* cells with no Ti plasmid, they can't transform a plant.

This was a kind of magic. A bacterium had the ability to genetically hijack plant cells by transferring a piece of DNA into the plant (Figure 2.8). This T-DNA contained a number of genes, and these genes were quite special. The coding regions of these genes encoded enzymes that made the plant cells divide in an uncontrolled manner, much like a cancer/tumour in a mammal.

Figure 2.8 Agrobacterium-mediated transformation of plants. The natural system transfers genes that cause crown gall disease. The engineered system removes the disease-causing T-DNA and replaces it with the genes we want to transfer to the plant.
Image credit, Lara-Simone Pretorius.

That was because the enzymes encoded on the T-DNA genes are actually biosynthetic pathways to make two very powerful plant hormones, auxin and cytokinin. Crucially there was another gene type on the T-DNA, a gene encoding an enzyme which made those same plant cells synthesize a previously little known conjugated amino acid (an amino acid + a simple sugar) called opines. These opines are a key to understanding how the plant–pathogen interaction works. Opines are quite rare in natural systems, but the first report of one was that isolated from octopus muscle in the 1920s and called octopine. As a result of this rarity, very few organisms have the ability to catabolise them (break them down as a food source). One of the few organisms that has this ability is *Agrobacterium tumefaciens*. Hence when the *Agrobacterium* transforms the plant cell, the overproduction of auxin and cytokinin makes a tumour, and inside the spaces between cells, opines are abundant, and hence this is a wonderful micro-environment for the bacterium to feed and multiply, as they have created a niche in which they have a selective advantage over most other organisms. Only the *Agrobacterium* cells have the ability to use opines as a source of energy (carbohydrate and nitrogen).

Noteworthy (and let's face it, biologically an absolute wonder) is the question as to what was the amazing quirk of nature that allowed *Agrobacterium* to accumulate the genes for whole pathways of plant hormone biosynthesis AND have these genes under the control of eukaryotic promoters. These genes will never be expressed in the bacterium, as their promoter sequences will only ever work in a plant cell.

After Chilton demonstrated the transfer of the T-DNA, the advances continued. One of the next crucial discoveries was the question of not so much what the T-DNA is, but how does the *Agrobacterium* gene transfer apparatus work out what indeed is the T-DNA? There must be some molecular signposts that determine what part of the Ti plasmid is actually transferred to the plant. And this breakthrough was made by Patricia Zambryski in the van Montagu lab in 1982. They showed that there are small inverted repeat DNA sequences at either end of the T-DNA, which they named border sequences. These sequences are very small, only 24–26 nucleotides, and the gene transfer apparatus (known as the virulence proteins) actually recognise these borders and make a copy of that DNA to transfer to the plant. Even more exciting was the fact that it is only those border sequences that the virulence proteins recognise – the genes between them are of no actual consequence – gene transfer will occur regardless. This finding was absolutely pivotal. Using "natural" T-DNA to transfer foreign genes to a plant had a major shortcoming. By transferring the oncogenes to the host plant, you would never be able to regenerate a whole plant, as the overproduction of auxin and cytokinin would always cause that tumorigenic or neoplastic growth of a crown gall. But within a year, Patty Zambryski was able to make a vector without the oncogenes and you could transfer any genetic material, provided it was within the border sequences, without altering the regenerability of the plant tissue. Bingo! Now to get a transformed plant with a piece of foreign DNA that actually changed the phenotype of the plant, *i.e.* a gene that under the right conditions gave the plant a selective advantage.

Finding himself in the midst of this race was a young Mexican PhD student, Luis Herrera-Estrella. Today, Luis works in CINVESTAV, a national polytechnic and research organisation in Mexico. Luis is at the Irapuato campus for Biotechnology and Biochemistry, where he is the Director of the National Laboratory for Genomics for Biodiversity, and is one of the nicest guys you could hope to meet. How he found himself in Ghent in the early 1980s was "a bit of an accident". He trained in Mexico as a biochemical engineer, and he studied molecular biology and genetics during his Master's degree. He was determined to use microbes to develop new antibiotics, which were expensive in Mexico, and he wanted to "do something good for Mexico". He was told by a few more senior people that this was mostly tied up by pharmaceutical companies, and besides, it was very costly to get through all the regulatory hurdles to make it to market. Meanwhile, a plant molecular biologist from Leiden in the Netherlands, Rob Schilperoort, visited Mexico. Schilperoort was working with van Montagu and Schell on aspects of crown gall disease, and he encouraged Luis to think about how this bacterium could transfer DNA to plants. Shortly following that, Luis attended a conference on plasmids in Jamaica, where he met

Marc van Montagu. In September 1981, Luis found himself in the van Montagu lab in Ghent, working on T-DNA border sequences with Patty Zambryski.

Within a few months he was making his own recombinant DNA (or as we know it, moving different LEGO blocks around), and was starting to do some transformations with *Agrobacterium* and tobacco.

> "I didn't really think that what I was doing was so import-ant, but I was having a lot of fun. I thought I must be the only person working on this. Then a lady from Austria came and visited [he thinks probably Marjori Matzke], and she told me that I was only one of the many people also working on getting genes into plants."

In November 1982, Jeff Schell came to see him in the lab, and asked him how he was getting along. Luis showed Jeff his tobacco callus tissues that were growing on kanamycin (they had the *nptII* gene) and also expressing another antibiotic resistance gene, the chloramphenicol acetyltransferase (CAT) gene. Schell told Luis that he had heard that an American company, Monsanto, was going to announce its success at getting the first transgenic tobacco plants with a foreign gene in January 1983 at a conference in Miami. As it turned out, Jeff Schell, Mary-Dell Chilton (now at Washington University, St Louis) and Robb Fraley from Monsanto all announced that they had transferred chimeric foreign genes to tobacco. When Schell returned from the conference and showed Luis the abstracts, it was a watershed moment. It was then that Luis realised "this was a big deal", and he raced to regenerate tobacco plants and assay them for kanamycin resistance and assay for CAT expression, so he could submit his paper in February/March 1983. And as it turned out, in 1983, all three groups published the successful regeneration of transgenic tobacco plants, and Luis Hererra-Estrella was the first author on two papers from the Schell/van Montagu group. The future of transgenic plants was born, so long as it was tobacco. What Luis Herrera-Estrella found most exciting was that here was a new technology that could be delivered in a seed. And that meant that it was a technology that could be delivered to all farmers, including the poor farmers of Mexico, and the farmers would not even need to change what they do.

"I never considered that companies could monopolise the technology, especially when Ghent and the Max Planck Institute had most of the patents."

So Luis was convinced (despite many offers) that he could have the most impact by returning to Mexico and working on the crop species most important to his home country.

"I had the possibility to make a real change – something completely new to make life better for Mexican agriculture. In addition, I'm not clever enough to keep at the top as a scientific leader in Europe."

We will find out more about what Luis has been up to since the 1980s in later chapters, and I think you will be left in no doubt that this guy is indeed "smart enough" to be at the top of the game wherever he is.

Soon after the excitement of 1983, there was a breakthrough in the development of direct gene transfer, without the use of *Agrobacterium*. Again it was in tobacco, but this time through the protoplast route. Sadly, for the Cocking group in Nottingham, they were gazumped by Ingo Potrykus's group in Basel, Switzerland, who regenerated the first transgenic tobacco without using "nature's genetic engineer" in 1985. In the same year, Rob Horsch from Monsanto published a study showing that using *Agrobacterium* they could regenerate transgenic plants of petunia and several cultivars of tomato. So there was the promise of new traits in many crops, not just tobacco.

But man does not live by tomatoes alone. And for certain you are not going to survive for very long on petunias or tobacco. And what humanity survives on all over the world is cereals. In fact, almost 50% of our caloric intake globally is provided by only three grass species: rice, wheat and maize. And it was an established fact that grasses (including all cereals) are not within the host range of *Agrobacterium*.

For the flowering land plants (angiosperms), plants can be divided into two major sub-groups, the dicotyledons and the monocotyledons. These represent a distinct divergence in plant evolution. We generally call the dicotyledonous plants the broadleaf plants, but botanically the name means they have two cotyledons or seed leaves in their seeds. Germinate a bean, pumpkin or tomato seed and those first two leaves that pop out are the cotyledons. And given that scientists, especially taxonomists, tend to be rather linear in their thinking, you're correct to surmise that the monocots have only one seed leaf. Grasses are monocots, as are quite a few important families including the lilies, the alliums (onions) and bananas. Monocots were deemed by a large study to be beyond the host range of *Agrobacterium*.

This simple fact has long since been debunked. For one thing, the Nester lab was able to demonstrate that the Ti plasmid determines the host range of any given strain of *Agrobacterium*. While it is true that *Agrobacterium* does not cause crown gall symptoms on most monocots, it doesn't mean that there is no T-DNA transfer. In many cases, it may be that the host plant tissues lack the ability to respond to the additional auxin and cytokinin concentration. Gunter Kahl's group in Frankfurt were able to show that if you exposed *Agrobacterium* to wounded dicot plant cells, you could get them to transfer T-DNA to yam cells (a monocot) and even form small crown galls. Yet cereal transformation remained out of reach. Given that the major three cereals provide half our calories and the fact that cereals are planted on half the world's arable land (wheat 16%, maize 13%, rice 12%, barley and sorghum 4% each), this was a problem.

And when you've got a problem, how do you solve it in the USA? With a gun of course. A collaboration between Cornell University and DuPont led by John Sanford and colleagues. To attempt to overcome the species barrier seemingly put up by the monocots, they decided that a direct approach at actually shooting genes into a plant may work. Admittedly it is in no way subtle compared to the natural elegance of *Agrobacterium*. The basic approach is to coat small inert metal particles such as tungsten or gold with the DNA you wish to transfer and shoot these micro-projectiles into plant tissue that you know you have a good chance of re-generating whole plants from. Nothing like a bit of brute force to make things happen.

The usual starting material may be plant pieces or, more commonly, callus tissue. These tissues are introduced into a special chamber (see picture), then placed under a slight vacuum and shot with the DNA-coated micro-projectiles, as many as one million of them in a single shot. Quite a few cells die in the process, and quite a lot more cells are missed by the shot. And of those cells which are hit, most of the time the DNA is not delivered to the right place, the nucleus. But, perhaps a few hundred or so of these micro-projectiles actually make it right into the nucleus, and in some cases, the DNA is then incorporated into the chromosomes and passed onto the next generation of cells. And because this happens at such low frequency, we need to give those cells a selective advantage. That's right, just like with those other methods we include an antibiotic resistance gene, and add the antibiotic to the tissue culture growth medium to kill off those cells that don't have our gene of interest. The first successful demonstration of this technique was with onion cells in 1987, so it worked on a monocot species. Now to make it happen with a cereal and preferably (for seed companies) with maize, their favourite commodity.

Peggy Lemaux is now a Professor at University of California Berkeley, and her official responsibilities are 50% research and 50% outreach – the public communication of GM crops. She is the Energizer Bunny of plant genetic transformation research, and communication of the science, and a wonderful person to discuss and "do" science with. By her own admission, back in the 1980s she knew nothing about plants or agriculture. She trained under Stan Cohen at Stanford and described herself as a good molecular biologist. She said:

> "Stan was a very deliberate person. He taught me how to solve giant puzzles. He taught me how to really put a project together and even more importantly, how to beat the other people to finding the answer first."

In 1987, Peggy found herself up in Mystic, Connecticut, at a small biotechnology lab run by the DeKalb Genetics Corporation. DeKalb was one of the oldest family seed businesses in the

USA, starting in 1912, but its biotech arm was a joint venture with Pfizer, the pharmaceutical company. Peggy's challenge was to help DeKalb recover its market share in the competitive maize hybrid seed market, which had fallen from 23% to less than 10%. And one way forward was to develop a method to make transgenic corn, a hugely competitive area with the big hitters like Pioneer Hi-Bred and Monsanto pouring millions of dollars into their programmes.

> "Ingo Potrykus (now at ETH Zurich) said that nobody would be able to make transgenic cereals, and if they do it won't pass on to the next generation. We were truly the little guys, flying under the radar, and nobody in corn biotech knew who we were."

According to Peggy there were two big things that allowed them to make a breakthrough. There were some problems with the normal antibiotic selectable marker genes in maize, and Peggy had heard from a colleague about a new herbicide resistance gene, the *bar* gene, which conferred resistance to the bialaphos herbicide. This was a powerful herbicide that blocks the main enzyme responsible for turning inorganic nitrogen (the form plants take up from the soil) into organic nitrogen (amino acids, the building blocks for all proteins). The second decision made was about using the gene gun.

> "I had heard about the gene gun and wanted to try it. I went to our bosses and said I wanted to buy one of these guns. They pointed out that to buy a gun would use all of our research budget for an entire year. I described what it could do and they said "Really? You are going to shoot corn with little cannon balls?" It was a career-defining decision and I'm glad I made it. If it didn't work, it was the end of my career."

In 1990, Peggy was at a Gordon Conference in Asilomar on the California coast. They had made transgenic corn plants, but they needed to demonstrate that the genes had passed onto the next generation, and the only acceptable way to show the foreign DNA was correctly inherited and incorporated into the genome was to extract DNA from the progenies and do what was called a Southern blot.

> "The morning before my talk, I sat by the phone. I couldn't make an announcement before we had the Southern confirmation, and without the Southern confirmation we couldn't file a patent. We had the patent all ready to go and just needed to insert the Southern blot photo. So I was waiting for the lab back in Mystic to confirm. And then the phone rang ….and it was positive! I went along and gave my little seven-minute talk. Just a seven-minute talk to say we were the first team in the whole world to report transgenic corn! From a scientific standpoint, plant biology is magical – and we did magic!"

Micro-projectiles shattered the myth that the cereals couldn't be transformed. And shortly after Peggy's maize breakthrough, rice followed and by 1994 the other big cereals – barley, sorghum and wheat – had all been transformed. And in the same year, Yukoh Hiei and his team from Japan Tobacco had successfully transformed rice with *Agrobacterium*, demonstrating once and for all that the most important food crops, the cereals, were ready to join the agricultural biotech revolution.

REFERENCES

1. J. K. Rowling, Text of J. K. Rowling's commencement speech following the awarding of her honorary degree, *The Harvard Gazette*, Thursday June 5 2008, https://news.harvard.edu/gazette/story/2008/06/text-of-j-k-rowling-speech/.
2. J. B. Ristaino and D. Pfister, What a painfully interesting subject: Charles Darwin's studies of potato late blight, *Bioscience*, 2016, **66**, 1035–1045.

CHAPTER 3

Revolution

"An old jest runs to the effect that there are three degrees of comparison among liars. There are liars, there are outrageous liars, and there are scientific experts."[†]

And to paraphrase Mark Twain in *Chapters from my Autobiography*,[‡] the most contemptible of these experts would have to be the statistician. The statistician will always tell you that the numbers don't lie, but have you actually framed the question correctly?

Hopefully you're reading this book because you have a need to gain an understanding of the impact of genetic technologies on how crops are produced and what it means for you as a farmer or consumer. Or to put it another way, are GM crops good or bad? Spoiler alert! This chapter is mostly about the good, and as you are an educated reader you want to have a few facts. So you are obviously literate, and in all probability, also highly numerate. For illustrative purposes, it is now necessary to quote some statistics, or as Mark Twain would have us believe, those things

---

[†]Attributed to Sir Robert Giffen, a Scottish Statistician and Economist (*Economic Journal* 2(6) (1892), 209–238), although a similar phrase was popularised by Mark Twain in 1906 and incorrectly attributed to former British Prime Minister Disraeli. Twain, of course specifically chose to limit the field of scientific experts to "statistics".
[‡]Mark Twain (1906). Chapters from My Autobiography. North American Review.

Good Enough to Eat? Next Generation GM Crops
By Ian D. Godwin
© Ian D. Godwin 2019
Published by the Royal Society of Chemistry, www.rsc.org

that are considerably more nefarious than lies or damned lies. So now that you are primed, here come some numbers!

- 37% less pesticide usage
- 22% increase in crop yields
- 68% increase in farmer profits

Some pretty good numbers, if you want to make statements in support of the economic and environmental credentials of GM crop cultivation. The statistics are the outcome of the largest "meta-analysis" of the benefits of GM crops to date, carried out by Klümper and Qaim in 2014.[1] A meta-analysis is a study based on the analyses of a large body of other published work, and in this case, the team from Göttingen University in Germany painstakingly analysed 147 publications on the impact of genetically modified maize, soybean and cotton around the world. The first genetically modified (although we called them genetically engineered back then) crops had been commercially released in 1996, and by 2014, 181 million hectares was sown to GM crops, more than half of which was in developing countries. They chose not to include other crops such as canola/rapeseed, sugar beet and papaya because these crops were either too new or had not been grown on significant areas by that time.

Firstly, what is 181 million hectares? It's hard to conceive – maybe I should say that it's the size of 181 million football fields, but does that really help? Another way to put it is that it is 1.8 million square kilometres. So that's just slightly smaller than Mexico, or equivalent to France + Spain + Germany + Italy. So imagine if the entire land surface of those four European countries was covered in GM crops (Figure 3.1). And I mean the entire surface. So, we would have to get rid of Paris, Milan, Berlin, Barcelona and all those other wonderful places to put in GM crops. And many among Germany's population would die of apoplexy at the thought. Imagine replacing all those beautiful biergartens with their seriously pure beers and putting them under GM maize. Yes, that disturbs me too. Not least of which because the German Greenpeace activists would probably go crazy with scythes to rid their land of the GM scourge that they may end up accidently killing one another in a reaping-frenzy. Then they would have the first fatality ever caused by a GM crop.

Figure 3.1 Worldwide area of different GM crops as pictured in the equivalent land area of European countries. Note the SALAD of Switzerland represents the combined area of sugar beet, papaya, potato, carnation, eggplant (aubergine/brinjal), apples, poplars, squash, sweetcorn, lucerne/alfalfa and mustard. Areas are from the International Service for the Acquisition of Agri-Biotech Applications (isaaa.org). Image credit, Lara-Simone Pretorius.

So let's give Germany a GM moratorium and leave them out of the equation and replace them with the UK (which is in Europe despite what you Brexiteers would like to think), the Czech Republic and Switzerland.

So now if you totally cover (in your mind's eye): France and UK with soybean, Spain entirely with maize, Italy with cotton, Czech Republic with canola/rapeseed and give Switzerland (in every valley, mountain top, glacier and lake) a healthy organic mix of GM sugar beet, papaya, potato, carnations, eggplants, apples, poplars, squash, sweet corn, alfalfa/lucerne and mustard, well you have a rough picture of the land area currently covered by GM crops. I do feel sad about removing Prague from the map because my son tells me the beer is great and only €1 for a pint.

If you came from another planet and landed randomly in Australia, it is highly likely that you would find the place rather uninviting. That's because over 70% of the country is desert – or what geographers call arid landscapes. Much of the remainder is what we call semi-arid, meaning it doesn't rain much there either, and the daily evaporation far exceeds the precipitation that falls on the ground. It's hot, it's dry and it's inhospitable for plants and animals. For this reason, nearly everyone (plant or animal) clings to the coastal fringe. And that is also where most of the agricultural production takes place. So a vast country/continent like Australia only cultivates around 25 million hectares for food and fibre production on an annual basis and has around 47 million hectares of arable land. Agricultural powerhouse nations like China, India and the USA have 105, 156 and 154 million hectares of arable land, respectively. So if all the arable land in any one of those countries was under GM crop production, it would still not account for the total GM area.

Among the currently grown GM crops there are two traits and four species that dominate the numbers. The two traits are insect resistance and herbicide resistance, and the Fab Four crops are maize, soybean, cotton and canola/rapeseed. Surprisingly, these were not the first GM traits to be commercially released. That honour lies with the FlavrSavr tomato, of which we will no longer speak.

Cover it up, sweep it under the carpet, it didn't happen. Well OK, in the interests of full disclosure let's briefly recap on the sad story of the FlavrSavr tomato. Being a FlavrSavr tomato meant the tomato would ripen and stay a beautiful red, juicy tomato and not get damaged by being shipped around in a truck. It was a delayed ripening trait. Such an exciting moment for us gene jockeys when FlavrSavr tomatoes made it to a supermarket in your street. Sadly, however, FlavrSavr rapidly became a public relations disaster and caused the demise of one of the more innovative small biotech companies. It didn't make it to a supermarket in my street either, because it never made it out of the UK or USA, and in fact, when Friends of the Earth and Greenpeace got involved, not many actually made it out of the supermarket. So at the mention of FOE and the G-word, I've suddenly developed a nervous tic and hence I think we'll postpone this discussion until a later chapter.

So back to those first two traits, which yes, yes you are correct as always, are actually the second and third traits.

3.1   HERBICIDE RESISTANCE – ROUNDUP READY

The first trait to explode onto the market with the promise of better yields, better crop management and better profitability, was herbicide resistance. There are lots of herbicides around that do all sorts of things to plants. Some mess with their hormones, like the synthetic auxins. Some mess with the ability of the plant to photosynthesise, and some prevent a plant from turning the inorganic nitrate and ammonium it has taken up into organic amino acids. Others affect the ability of the plant to manufacture cellulose (essential for cell walls), or lipids (essential for cell membranes), or carotenoids (essential for pigments). Herbicides are big business, and it's not just farmers who want to control weeds. Local councils and governments need to control weeds in roads, roadsides and parkland, and home gardeners also use herbicides. In many cases it may be for cosmetic reasons but this isn't usually the case in on-farm situations. Weeds in crops compete with the plant for water and nutrients, and in some cases when they are taller than the crop, they actually compete for sunlight. They may also be sources of habitat for the increase in pests and disease, and some may even produce toxic seeds, which you definitely do not want to get mixed into your harvested crop. Weed seed contamination in crops is a major international trading issue. Countries like Japan regularly reject shipments of grain or hay from Australia, Canada and the USA because of weed seed contamination.

In the 1970s, Monsanto had developed and commercialised an absolutely fabulous broad-spectrum herbicide which was gaining a lot of usage worldwide. That herbicide is sold under the name Roundup, and its active ingredient is glyphosate. It's an absolute cracker of a herbicide because it attacks a particular pathway in plants that does not exist in most other organisms. It is a systemic herbicide such that when glyphosate is applied to any part of the plant (by spray, paint brush, weed wand or even spray drift), the plant loses the ability to make three crucial amino acids, and within 10 days or so has run out of these amino acids and it dies. You will recall that amino acids are the building blocks of proteins, and hence if a cell cannot make some of the required amino acids, it has to get them from somewhere. I'm going to dive a bit deeper into the biochemistry and molecular biology of how the herbicide works so if you're not so interested, skip a few paragraphs and join back in over the next page.

Glyphosate targets a single enzyme, which you will be absolutely beside yourself to know is called 5-enolylpyruvalshikimate-3-phosphate synthase. We call it EPSPS because we can't remember how to spell it either. Being a synthase, it synthesises something, and in this case that something is EPSP. It is part of the shikimate pathway and the end point of this pathway is a group of amino acids known as the aromatic amino acids – phenylalanine, tyrosine and tryptophan. They are not aromatic in the olfactory sense, like asparagine which is the aroma of asparagus, but because they have a benzene ring central to their structure. We humans, like most animals, cannot make phenylalanine or tryptophan, so these enzymes are essential parts of our diets. Fortunately, all plants and microbes can make these amino acids and so we need to eat plants to get the amino acids. That's a double attraction of EPSPS as a herbicide target: (i) it targets an enzyme that is essential for plants to make three key amino acids, which are required to build proteins; and (ii) that enzyme target is non-existent in animals so the herbicide has no effect on our ability to make proteins.

A bigger question is: What possible genetic approaches are there to render a plant resistant to a herbicide? Remember that this herbicide is a toxin which binds to an essential enzyme to prevent it from being active. Nature has found three ways to do this with regard to any toxin. These are:

1. Detoxify the toxin before it does any harm.
2. Overproduce the target enzyme so that there is still plenty of active enzyme.
3. Produce a version (isoform) of the enzyme to which the herbicide cannot bind.

And in this case the winner was number 3. Another of the advantageous features of glyphosate it that after application it is rapidly degraded in the soil. That showed strong evidence that some members of the soil microbiota were capable of using glyphosate as a food source. It also showed that because these members of the soil microbial population were not killed by the glyphosate, they must be resistant to this toxin. So, Sherlock, I hope you've worked out that these microbes are a good source of potential resistance genes. And it turned out that the microbe which became the source of resistance was the one we spoke of in the last chapter: *Agrobacterium tumefaciens*. An *Agrobacterium* strain was found to contain an EPSP isoform which was totally resistant to the herbicide. So now we have the middle LEGO block to the gene – the coding region. There was still lots to do to ensure the gene would express at the right levels and in the right tissues.

Ideally, to make a plant highly resistant to the herbicide, you would like the resistant isoform of the target enzyme to be expressed in all tissues at all times (constitutive expression). In many plants, such as tobacco, one of the best promoter sequences for high-level constitutive expression is of viral origin, from the cauliflower mosaic virus 35S promoter (CaMV 35S). Scientists were even able to enhance the expression from this promoter, by duplicating the enhancer region of the promoter, creating what became known as double 35S or enhanced 35S. There is some dispute as to who made this discovery, with the University of British Columbia and Monsanto fighting over the intellectual property. Why is it that Monsanto is always one of the parties disputing who "owns" this stuff? I'm going to defer that discussion until Chapter 4, *Chemical Heart*.

And of course, a terminator LEGO block was needed to round off the gene and the *nos* terminator from the nopaline synthase gene of *Agrobacterium* was ideal for the job.

The enhanced 35S promoter was good at boosting transcription of the gene, but there were also some other ways to ensure the enzyme was stable for a long period of time. To really make large amounts of the enzyme, it wasn't just about boosting transcription. If you can target the final protein to a nice cellular compartment where it can do its work without being constantly degraded, then why not go for that as a post-translational modification?

One of the first available means to do this was to use a transit peptide. A transit peptide ensures that a newly minted polypeptide ends up in exactly the right place. One of the first such systems to be elegantly elucidated was the means by which the key enzyme in photosynthesis is put together by the gene expression and translational machinery of photosynthetic cells. It had been originally proposed that RuBisCO, the key enzyme of photosynthesis, required two different sub-units. An active RuBisCO molecule needs eight small and eight large sub-units which are assembled in the chloroplast. Intriguingly, the large sub-unit is encoded on the chloroplast genome. However, the numerous genes for the small sub-unit are all on the nuclear-encoded genes (on chromosomes).

So how does a protein from a gene encoded in the nucleus, transcribed in the nucleus, and translated in the cytoplasm of a cell, get transported accurately and reliably to the chloroplast? The team at Ghent were able to show in 1984 that the presence of a specific transit peptide was enough to transport the RuBisCO small sub-unit to the chloroplast. So in this sense, the transit peptide is not part of the active enzyme, but is cleaved off the remainder of the small sub-unit and recycled. Plant cells are clean and green, especially the photosynthetically active ones.

This knowledge of the RuBisCO small sub-unit was then harnessed by Monsanto to deliver the translated peptide with the resistant isoform of EPSPS to the chloroplast using a signal peptide from petunia. Hence the LEGO block models were getting more complex – to deliver a high level of herbicide tolerance the full gene required five blocks, two (duplicated) from a virus, a signal peptide from petunia, a coding region from *Agrobacterium* and a terminator from the same species. Of course, what is immediately evident to many of you was that most of the DNA delivered to achieve herbicide resistance was not of plant origin.

So the Roundup Ready soy went into commercial production in 1996 and let's just say it would have to be regarded as a success story. Before I go and smack you around the head with loads more figures, let's get a bit of a picture of the soybean industry before Roundup Ready or RR soybean. You may not know too much about soybean beyond the fact that it is used to make tofu, cooking oil and soy sauce, a fermented product extremely popular throughout eastern Asia. You may also not know that in their raw form they produce trypsin inhibitors. Trypsins are enzymes we produce in our gut to break down proteins, hence if we eat raw soybean they are actually toxic in the true sense. That's why we always cook them and/or ferment them to break down the trypsin inhibitors. Imagine, one of the top 10 most important food crops and it's toxic. Sweet potato also has trypsin inhibitors but so long as you cook them these are denatured. Other top 10 food crops with significant life-threatening toxins include cassava (cyanide), potato (alkaloid), sorghum (cyanide but not in the grain), wheat (gluten). Hang on, gluten is not a toxin. Have I already said that? Well here it is again. Gluten is not a toxin.

And then there's the banana story. Bananas are rich in potassium. Potassium is an essential nutrient for the maintenance of all sorts of cellular water balances, just little things like making sure your heart beats correctly. Too much or too little potassium will have potentially life-threatening effects on your heartbeat. There was one of those urban myths going around a few years ago that if you ate more than six bananas you would ingest so much potassium that you heart would actually stop. Another great subject to bring up at the Sunday barbecue, and then some bloke (and it's rarely a female) who has fuelled up on beer since 11 am decides to eat seven bananas just to test the theory. And the internet is full of such stories. Catherine Collins, from St George Hospital in London, told the BBC in an interview in 2015 that you would actually have to eat about 400 bananas to approach a toxic level. So there it is, that whole reality check thing about toxicity. Nearly everything is toxic, including most of the things that are essential for us. If we have too little (or none) it is a deficiency problem, and if we have too much, it's a toxicity problem. Of course, some things are just toxic. We don't need a single molecule of them.

OK so back to the soy story. It's not just in tofu or your soy milk. In fact, these make up a very small part of the market for soy. Soy is a major bio-industrial crop with many end-uses. In most cases, the soy is crushed and the oil is extracted as a source of vegetable oil for cooking and food preparations. The remaining meal is a great source of highly nutritious protein and is a major food source for intensive animal production, including for pigs and poultry. Over half the US soybean production is fed to chickens for both meat and egg production. But the oil and protein are used for a huge variety of end-uses. The oil is a pretty good substitute for many petrochemicals, and can be converted to bio-diesel. It is used in many automotive lubricants, and if you bought a Ford recently, maybe the foam in your car seats was made with soybean. If you are still so old-fashioned as to read a hard copy of the *New York Times* or *USA Today* then much of what you are reading is printed with soy-based inks of many colours. Have a look at your chocolate. You will see on the wrapper that nearly all of it contains lecithin, which is a soybean extract that helps to emulsify the chocolate ingredients. And that's a pretty good thing because who wants to eat a chocolate where the cocoa and the cocoa butter have separated? And lecithin is also used in the manufacture of coatings and paints.

The biggest producers of soybean are the USA, Brazil and Argentina, who together grow about 80% of the world's production. Brazil is rivalling the USA to become the world's largest soybean producer. But it wasn't always that way. When RR soybean first came to South America, Argentinian farmers led the way. Greenpeace became very active in a campaign to stop the growth of RR soy and the Brazilian government announced a moratorium in 1998. However, when they could see that farmers just across the border seemed to be doing fine, and probably even better, then the temptation to try the new technology became too much for many Brazilian farmers. By 2003, it was publicly obvious that many farmers had ignored the government. Soybean seed was regularly smuggled over the border with Argentina, especially in the most southerly Brazilian state, Rio Grande do Sul. Rio Grande do Sul shares borders with Argentina in the west and Uruguay in the south, and soybean was smuggled in from both of those countries, such that by 2003, 90% of the soybean grown in the state was RR soy. The farmers were very happy because not only had their profitability improved markedly, but because they were growing the crops illegally, they didn't have to pay royalties on the seed or the harvested product. Who wasn't happy? Monsanto, and US farmers who did have to pay royalties, and felt that Brazilian growers were gaining unfair advantage. To add insult to injury, because officially there was no GM soybean in Brazil, some Brazilian soybean was being sold on international markets at a premium as organic "GMO-free" soy.

3.2　INSECT RESISTANCE – BT COTTON AND MAIZE

Weeds are of course not the only problem in crop production. Insect pests are serious threats to food security and no more so than the insect magnets known as *Gossypium hirsutum* and *Gossypium barbadense*, known more commonly as cotton. Cotton is a relatively "new" crop, and as a natural fibre was not particularly well known in many parts of the world, where linen or silk held sway among the fashionista and the general public alike. In some ways it has quite a sorry history, and although I learnt at school that what paved the way for the growth of cotton clothing and bedding was the invention of the cotton gin by Eli Whitney in the USA in the late 18th century. Well yes, he did get the first patent, but the Indians and Chinese had cotton gins all over the place more than 400 years before that. They just didn't have patent offices.

The cotton industry was founded on slavery, or some form of indentured labour, because the labour-intensive tasks of weeding and harvesting were all performed by hard labour. And even when mechanisation became available, the tasks of controlling weeds and insects were still major constraints to the profitability of the industry. Cotton is a prime example of what happens when you grow a crop as a monoculture – it becomes a magnet for pests and diseases. Cotton definitely attracts the wrong sort of crowd. And by a country mile, the biggest, baddest member of this crowd of loathsome individuals is the cotton boll weevil, or *Helicoverpa* (formerly *Heliothis*), which is a little brown moth about the size of your thumbnail. This moth is not what does the damage, it's the baby moths, or very hungry caterpillars. The moth is attracted to the cotton plants and lays eggs on the leaves and developing bolls. And when these eggs hatch, a voracious little caterpillar is on its way to eating as much cotton plant as it can, before it then undergoes the metamorphosis into the flying version of the species.

In Australia, the cotton bollworm (*H. armigera*) is only one of the big cotton pests, and the other is a close relative, the native budworm (*H. punctigera*). Both are important pests, and as a result, cotton farmers needed to spray their crops with a number of different insecticides to control them. Cotton became the biggest user of pesticides in Australia, especially as the bollworm started to develop resistance. The bollworm survives winter in the soil and re-emerges in spring. This is a problem, because the same populations sprayed with and surviving last year's insecticide applications emerge in the same place the following year. Although the budworm was just as voracious, it was migratory and spent the winter in the dry areas of central Australia where they were not exposed to the insecticides. So the budworm was easier to control, whereas the bollworm was in an evolutionary battle to develop resistance to whatever chemicals were thrown at them. So by the 1990s, Australian cotton growers were spraying almost on a weekly basis, and many were spending $1000 per hectare on insecticides! One of the most commonly used was endosulfan, a type of organochlorine that could kill the bollworm and the budworm. It could also kill mites, whiteflies, leafhopper, beetles – and you. And even worse, it has a long residual time, building up in waterways, and in foods. So as well as being toxic, it has some other little nasty surprises, like imitating the effect of oestrogen, so it can disrupt your endocrine system and have major impacts on reproductive ability. Can you imagine living in the main cotton growing areas of Australia, where every day, a crop duster would fly over dumping huge amounts of endosulfan and other goodies on the crop – and sometimes not only on the crop?

In 1998, 45% of sites tested in the Namoi, Gwydir and Macintyre river valleys were shown to contain endosulfan. And in the same year, a beef shipment to Japan was rejected because endosulfan was detected in the meat, probably from the use of cottonseed meal as an animal feed. Similar problems were being experienced in other major cotton-producing areas in the USA, India and China. There was much talk about banning endosulfan, and it finally happened in Australia in 2010. Cotton growers, their families and everyone who lived in towns like Gunnedah, Moree, Narrabri and Wee Waa were looking for a solution, possibly even the ultimate solution – not growing cotton anymore. Or growing cotton without insecticides and taking the hit of cotton yields reduced by more than half. One exciting and attractive solution lay in a soil bacterium, *Bacillus thuringiensis*.

Bacillus thuringiensis, or Bt as it is more commonly known, was first found and described in Thuringia in Germany in 1911. Ha!

Wrong again Eurocentric white boy! OK, it was actually discovered 10 years earlier in Japan by Shigetana Ishiwata, who discovered it was causing a disease in silkworms (silkworm good, bacterium bad). Then in 1911, Ernst Berliner discovered the same thing, causing a disease in the moths that get into flour (moth bad, bacterium good), and proposed it may be a potential control for these moths. Turns out Ernst was not wrong, and farmers started to use the bacterium as an insecticide in the 1920s, first made commercially available as Sporeiner in France in 1938. And then it turns out that Ernst was a bit wrong, because it wasn't actually the bacterium directly killing the caterpillars. By the mid-1950s, scientists in Canada had determined that the bacterium actually produces small crystals. These crystals were shown to be made of protein and it was these proteins that were killing the caterpillars. The proteins became known as Cry (for crystalline) proteins, and in many parts of the world, freeze-dried cultures of the bacteria were being sold as commercial products to control insects. Chemists all over the world got themselves involved in bio-discovery of new and more potent Bt toxins, with classifications going from Cry1, Cry2, Cry3 .... Cry67 and beyond. And within Cry1 there were sub-groups like Cry1Ab, *etc.* What we do know is that there are different host ranges within the different types of Cry proteins, but as a general rule the Cry1 and Cry2 target lepidopterans (moths and butterflies), Cry3 target coleopterans (beetles), Cry4 target dipterans (mosquitoes and flies), Cry5 and Cry6 target nematodes (which are not even insects). What an endless source of weaponry to target every known pest on the planet!

And the best part? When the crystals are ingested by the insect, they actually contain nothing toxic, but contain a pro-toxin. The enzymes and alkaline pH of the insect stomach cut the pro-toxin, leaving a toxic component which rapidly reduces the appetite of the herbivore, which then binds to specific cells in the gut of the insect and forms pores. What ensues is a not very nice death as the contents of the stomach leak out into the body resulting in death. Another real advantage of Bt toxins is that with a bit of careful selection, you can use one that has a narrow host range (meaning it may only be toxic to the boll weevil) and doesn't wipe out all other insects, leaving some of the beneficial ones to do some integrated pest management of their own.

But there were a few problems, not least of which was that the toxic proteins have to be ingested by the insect, and that when sitting on a leaf in the sun, they lose their toxicity (efficacy) within a few hours to a few days in most cases. And even if you spray with the actual bacterium, it gets affected by UV radiation (sunlight), or washed off by rainfall. So while it is true that Bt toxins are much more environmentally friendly than endosulfan and its ilk, you still have to spray every week, and that is the major part of the cost, the need for labour, and the carbon footprint of repeat doses of diesel for the tractor or aviation gas for the crop-dusting plane.

And this is where another soil-borne bacterium comes to the rescue – *Agrobacterium tumefaciens*, nature's genetic engineer. Why not just get the plant to make Bt toxin? You can get the gene into an *Agrobacterium* and let it do its magic, as discussed in Chapter 2. Then you don't have to spray every week, and the plants will be auto-protected.

The first GM cotton plants were released for commercial agriculture in 1996. In Australia, the project was a collaboration involving CSIRO, Monsanto and the cotton seed company Delta Pineland, where around 1500 families were growing cotton and becoming increasingly frustrated with the ever more severe effects of the caterpillars becoming resistant to their insecticides. The first Bt cotton was known as Ingard in Australia, although cottons using the same Cry1Ac gene were also released in the USA under the name Bollgard.

To say the first releases were straightforward and widely accepted would be to gild the lily somewhat. CSIRO scientists (rightly) were very concerned that the cotton only had a single Bt gene, and their population geneticists felt strongly that to release this line could potentially result in widespread resistance developing among the *Helicoverpa* populations, particularly among the *H. armigera* populations that had already become highly resistant to chemicals such as endosulfan. So farmers who did grow Ingard cotton were required to sign fairly restrictive contracts before they were allowed seed. The control measures came across as draconian to some farmers, and were also a boon to local anti-GM activists who wanted to drive a wedge between growers and the commercial companies. They may well have succeeded, except for the high regard that CSIRO agricultural scientists enjoyed among the rural community, and the

Australian community as a whole. A recent study by science communicator and social scientist Craig Cormick revealed that among the general public, CSIRO was the most trusted science organisation, and only pipped at the post by the Red Cross as Australia's most trusted organisation.

When Ingard cotton was released, the strategy was to allow no more than 10% of all cotton to be Bt in the first year, increasing to a maximum of 30%. Cotton growers had to adhere to the following rules:

1. They had to plant a "refugia" of conventional cotton of not less than 10% of total area, and leave this cotton unsprayed. They could grow refugia of maize, sorghum or pigeon pea (all hosts of *Helicoverpa*), but these areas had to be up to 20% of the total cotton area.
2. They had to plant all Bt cotton within a fairly narrow planting window, so that the insects were not exposed to the Bt toxin for the entire growing season.
3. They had to follow instructions on soil cultivation post-harvest, in an attempt to greatly reduce the boll weevil over-wintering population.
4. Insecticides could be used if necessary, but they could not use a Bt insecticide product.
5. Any Bt cotton volunteers had to be removed.
6. They could not keep cotton seed to plant next year, nor could they pass any cotton seed onto neighbours.

How aggressive were CSIRO and Monsanto in ensuring these rules were adhered to? Very! Although as one of my favourite lecturers, the late Professor Graeme Wilson, taught me as an undergraduate, "Every time you feel the urge to use 'very' in scientific writing, replace it with 'bloody' and you will soon come to recognise its redundancy". And I experienced some of this up close and personal. In the early 1990s, I proposed that we would use Bt toxin genes to develop sorghums that were resistant to *Helicoverpa* and another class of pest, the sorghum midge (*Contarinia sorghicola*). Very soon after, I was on the receiving end of phone calls and personal visits from one of the CSIRO scientists involved in the development of Ingard and its use on farms, Gary Fitt, who is now the Program Leader of Plant

Biosecurity at CSIRO. I was told, in no uncertain terms, that there was no way I could use Bt genes in sorghum, given that they were hosts of the same major pests as cotton. Deploying Bt genes in sorghum would make resistance management that extra bit harder, particularly as sorghum was grown in exactly the same geographic regions as cotton, and had a wide planting window from early September to late January.

Then finally, in 2003, Bollgard II was released – cotton with two different Bt genes, Cry1Ac and Cry2A. Population genetic theory tells us that the chance of a single insect simultaneously developing resistance to both of these toxins is pretty close to zero, although as anybody with a little knowledge will be quick to work out, the population was already exposed to Cry1Ac from 1996 to 2002, very possibly already enhancing the likelihood that some individuals would only need to then mutate to develop resistance to the new Bt toxin. However, within a few short years, almost 100% of Australian cotton was Bollgard II.

According to the peak body representing Australia, Cotton Australia, the benefits have been clear. Some growers reduced their insecticide usage to zero, with average reduction of 85%. This has of course led to massive reductions of costs, including labour and fuel, and corresponding increases in human safety, and populations of beneficial insects. By 2001/02, the number of sites still testing positive for endosulfan contamination was less than 5%. Within 4 years of the release of Bollgard II, cotton farmers were enjoying an average net benefit of $180 more farm income per hectare. And for the non-farm population in the cotton towns, complaints of insecticide spraying received by local governments had reduced from 127 in 1998/99 to only 10 by 2002/03. A quick look at the Australian government regulator of GM crops and medicines shows that of the (at the time of writing) 155 licence applications for trials or commercial release include 52 specifically for cotton, far in excess of the 18 for wheat and 16 for canola/rapeseed.

Very similar benefits were being reported by US farmers growing Bollgard and Bollgard II cottons. In the USA, prior to the release of these improved insect-resistant cottons, cotton was the second biggest user of pesticides (second only to potato, but more on that later). In 1995, 86 million pounds of active ingredient was applied to cotton fields. This had dropped to

37 million pounds by 2008. Remember, this is total pesticides, and includes herbicides and fungicides as well as insecticides.

3.3   BT MAIZE

At the same time, Bt maize/corn was released in 1996, the first generation containing a single Bt gene for the European corn borer, a relative of the cotton boll weevil. Although not as widely and uniformly devastating as the cotton pest, the European corn borer was nevertheless a formidable caterpillar. Throughout the 1970s and 1980s, insecticides for corn borer and other pests were applied to over 40% of US maize fields. Once Bt products became available for European corn borer and other major pests such as the root worm (another caterpillar), insecticide was applied to less than 20% of all corn by 2006. During that period however, herbicide use had increased to the extent that a maize field that had not had at least one herbicide application was a rarity. This had been the case well before the availability of Roundup Ready maize. Over 95% of US maize fields were herbicide treated throughout the 1980s, over 10 years before the first GM maize products were released.

Within a few years, Bt maize was being grown in many parts of the world, and as we will discuss in Chapter 5, it was not without controversy. Bt maize was the crop that really got the anti-GM activists active, and where they were able to gain some traction in getting the public interested in these "Frankenfoods", not just in the USA, the home of corn, but crucially in Europe.

3.4   DISEASE RESISTANCE – THE VIRUS-RESISTANT RAINBOW PAPAYA

I have a research collaborator at the university up the road, the Queensland University of Technology. Professor Rob Harding is a virologist, and has spent much of his career working on tropical fruits and vegetables as part of Australian government-funded agricultural research and developmental aid programmes. I was lucky enough to be involved in a project on germplasm collection and exchange of Pacific Island aroid root crops, mostly known as taro (or cocoyam). Taro is so important to many of the Polynesian and Melanesian cultures of the South

Figure 3.2 Samoan $10 bank note depicting the harvesting of banana, with taro growing under the banana plant (photo by Aleni Uelese).

Pacific that it makes it onto Samoan banknotes (Figure 3.2). In Vanuatu, local legend has it that humans came to be in existence on the islands after a brief romantic interlude involving a taro plant and a large monitor lizard. Just as plausible as Eve being made from one of Adam's ribs, so I will not venture into further biological discussion. Accept the diversity of belief and move on, dear reader. So back to Rob the much-travelled virologist.

Our project was a joint Australia–New Zealand Secretariat of the Pacific Community project on collecting, characterising and distributing taro germplasm around the Pacific. The project arose from an emergency in the early 1990s, when the devastating fungus, taro leaf blight, hit Samoa. Almost all the taro grown in Samoa at the time, favoured for its cooking and eating qualities, was called Niue, and the human disaster arose because it is highly susceptible to taro leaf blight, caused by *Phytophthora colocasiae*, a close relative of the *Phytophthora* causing late blight in potatoes. Taro was not just a staple crop for the Samoan population – it was THE staple crop, eaten by most people on a daily basis and the major source of carbohydrate calories. There were no sources of resistance to the disease within the relatively tiny taro gene pool available to the Samoan population. However, taro leaf blight had been in South East Asia, and the closer Melanesian island nations of Papua New Guinea and Solomon Islands, so there were known sources of resistant taros. BUT, these places also had some devastating taro viruses that did not exist in Samoa. Hence it was of paramount importance to transfer virus-free taro planting materials – let's not solve one problem (the fungus) while creating another (by introducing a whole complex of devastating viruses).

So let's be honest – Rob is a weirdo. Like most plant pathologists or entomologists, he loves nothing better than seeing a sick or infested plant. Tropical fruits and vegetables are some of the richest sources of a whole slew of pests and diseases. The tropics are a pretty tough place to grow plants. In the wet tropics particularly, perhaps because they can produce so much delicious biomass, plants have developed lots of enemies, from the microscopic viruses, phytoplasmas, fungi and bacteria, to insects large and small. And because the tropics have not been as well studied as most of the temperate world, that's where you are most likely to discover new organisms. *Sensu stricto*, viruses are not organisms because they lack the ability to reproduce – they need to invade a host and hijack the host's complexity to enable themselves to replicate. So viruses are really just encapsulated (usually in protein) DNA or RNA, encoding very few genes. Rob loves new projects in the tropics, not only because he gets to see a lot of sick plants. He also gets to use molecular biology techniques to characterise the nasty viruses lurking in plant tissue, and in the process of looking for the known ones, there exists the delicious prospect that he may discover some hitherto unknown little micro-beasties.

The viruses we were most concerned about in taro were the devastating Alomae–Bobone viruses. In local Melanesian languages, Alomae–Bobone means something along the lines of "dead and smells like shit", so these were some relatively easy viral symptoms to identify, even for me. However, the disease was actually caused by two viruses (large and small). When only one of the viruses was present, there were generally asymptomatic. Hence the prospect of introducing both viruses to Samoa individually in seemingly virus-free plants was not an unlikely scenario. Then once in Samoa, all that was required was for an insect vector to transfer the Alomae virus into a plant already infected with the Bobone virus, and the taro fields of Samoa would rapidly descend into miasma, and Samoa would have another food security problem.

So Rob and his team had a lot of fun developing molecular techniques using the polymerase chain reaction (PCR) to rapidly screen taro lines from all over Melanesia, Micronesia and Polynesia for these viruses. Once plants were deemed to be virus-free, they could be multiplied using plant tissue culture, stored in germplasm banks in Fiji and Papua New Guinea, and sent anywhere in the world (including Samoa) as virus-free plant materials. And along the way, Rob got to identify a few novel viruses from taro.

Because viruses are very simple pieces of nucleic acid, they are readily and rapidly mutable. As a result, viruses have a habit of jumping into new species, and creating new problems. You are of course acutely aware of this phenomenon, most famously from the Spanish influenza epidemic that swept the world in 1918–20 when it jumped from pigs to humans, and HIV, which seems to have come from another primate in Africa into humans in the 1960s.

Among the countless tropical fruits that have succumbed to new viruses is the papaya. Hawaiian papaya growers had their industry devastated by a new virus in the 1950s. The entire plant looked like it had come down with chicken pox. This virus caused plant defoliation, which resulted in severe reduction of fruit yield, sometimes to as low as zero, and subsequent fruit death. Even if fruit were set, these were usually significantly smaller than what was considered marketable size, and worse still, the skin was pock-marked with ugly ringspots. Hence the virus was named papaya ringspot virus (PRSV). To overcome the scourge of the virus, Hawaiian papaya production was moved from the island of Oahu to the "big island" Hawaii, and quarantine restrictions were enforced to exclude the virus from the new production location. But, as is the way of the brutal world of agriculture, "you can run, but you can't hide", and the virus finally caught up with its forcibly transmigrated host. PRSV symptoms started to appear in Hawaii in the 1990s, and this time, the virus seemed even more virulent. The virus was transmitted by aphids and within 3 years, the Hawaiian papaya industry was brought to its knees.

Like the taro situation with leaf blight, geneticists and pathologists teamed up to find a source of resistance. Unlike the taro situation, there was none to be found within the species *Carica papaya*. However, some wild south east Asian relatives such as *Carica cauliflora* were found to be resistant. A clever young PhD student in my lab, Pablito Magdalita, a plant breeder from the University of the Philippines, Los Banos, set to work to make interspecific crosses. The species were not readily interfertile, and he had to develop a technique known as embryo rescue to harvest interspecific embryos from the crosses before they aborted. He then had to nurture these crosses to maturity and harvest seed from them for further backcrossing to the papaya parent to get edible progenies. However, the species were so incompatible that they never produced any fertile offspring. The barriers between the species were insurmountable. Hence it was a *primae facie* case to look to the newly developed technique of genetic engineering to try and develop virus-resistant papayas.

As already mentioned, viruses are small pieces of nucleic acid surrounded by a capsid, made up in most cases of a single protein, also known as the coat protein. Roger Beachy (now at the Donald Danforth Centre in St Louis) and his team at Monsanto had shown that by expressing the coat protein gene from the tobacco mosaic virus in a tobacco plant, you could render the plant totally immune to the virus. A scientist at Cornell University, Dennis Gonsalves, wondered if the same methodology could be used to produce papaya plants resistant to PRSV. After all, the virus was in the same viral family as TMV, the potyviruses. Gonsalves teamed up with one of the developers of the "gene gun", John Sanford, and a scientist from Upjohn, Jerry Slighton, who had already cloned the PRSV coat protein gene. With a graduate student from the University of Hawaii, Maureen Fitch, they were eventually able to produce nine transgenic lines with the coat protein gene, six in the papaya cultivar Sunset, and three in Kapoho. In 1995 they were granted permission to grow some of the transgenic papayas in the infected region of Hawaii. After 27 months in the field, not one single transgenic papaya showed any of the tell-tale ringspot symptoms, whereas over half the non-transgenic controls were infected. One of the lines, known as UH Rainbow, averaged 100 000 lbs of fruit per acre per year. The non-transgenic lines produced only 5000. By 1997, the Food and Drug Administration (FDA) had approved the safety of the fruit for human consumption, and the Hawaiian papaya industry was back on its feet.

As a little side note, the virus also spread throughout much of the South Pacific and I was able to witness the devastation that the virus caused. In gardens where most of the fruit production was bananas, guavas and papayas, the papayas were rapidly disappearing. Those papayas that were in the market were small and covered in spots. In 2003, when I was leading a workshop at the University of the South Pacific in Fiji, we visited some fruit markets, and were surprised to see large papayas with unblemished skins for sale. And they were delicious too. When I asked where these came from, I was given the Pacific Island shrug, but then told they came from Hawaii. Somebody had brought in some seeds from Hawaii, and shared them around and everybody was happy that they produced big delicious papayas with no signs of the devastating PRSV. Regulations? What regulations? Here was another example of a farmer-led revolution in food security and maintaining the traditional diversity of diets.

BIBLIOGRAPHY

Global Status of Commercialized GM/Biotech Crops: 2017, Brief 53, ISAAA, *[last accessed April 2018]*.

www.tradingeconomics.com, a collation of national data on trade, economics, and production, *[last accessed April 2018]*.

M. Wishnick, M. D. Lane and M. C. Scrutton, The interaction of metal ions with ribulose 1,5-diphosphate carboxylase from spinach, *J. Biol. Chem.*, 1970, **245**, 4939–4947.

L. V. Herrera-Estrella, Nature, 1984, **310**, 115–120.

D. Gonsalves, S. Tripathi, J. B. Carr and J. Y. Suzuki, Papaya Ringspot virus, *The Plant Health Instr.*, 2010, DOI: 10.1094/PHI-I-2010-1004-01.

Can eating more than six bananas at once kill you?, David Rhodes, BBC News Online, 13 September 2015, *[last accessed April 2018]*.

Brazil's Controversial Cash Crop: Illegal Genetically Modified Soy Bean is Boon to Farmers, Utah Daily Herald, 21 December 2003, *[last accessed April 2018]*.

C. L. Hannay and P. Fitz-James, The protein crystals of Bacillus thuringiensis, *Can. J. Microbiol.*, 1955, **1**(8), 694–710.

REFERENCE

1. W. Klümper and M. Qaim, A Meta-Analysis of the Impacts of Genetically Modified Crops, *PLoS One*, 2014, **9**(11), e111629.

CHAPTER 4

Chemical Heart

"The country which is in advance of the rest of the world in chemistry will also be foremost in wealth and in general prosperity."

William Ramsay, Scottish chemist

Scientists and farmers were excited about all the new agri-biotech developments in understanding and changing the way plants could work for us. So were the corporations, who thought they could actually get on the bandwagon and harness some of the new genetics to increase their market share and asymptote towards world domination. Well, that's not strictly fair. Some were really interested in developing products that could benefit farmers and consumers (to make money). Some were really interested in developing products that could benefit the environment and the agro-ecology both on- and off-farm (to make money). Some were interested in ways that they could make tobacco with higher concentrations of nicotine, so they could increase the addictive nature of tobacco. The latter group is worthy of our general contempt. They funded research to increase the level of nicotine in the 1980s, while as late as 1994, their CEOs (of the seven largest US tobacco corporations) were testifying at US Congressional hearings that they "believed" nicotine was not addictive to humans (you can see it on YouTube). How do I know

Good Enough to Eat? Next Generation GM Crops
By Ian D. Godwin
© Ian D. Godwin 2019
Published by the Royal Society of Chemistry, www.rsc.org

this? I was working at the University of Birmingham in the 1980s and there was a team in the same lab doing just that, with financial backing from a British tobacco company.

Now it's true that corporations do exist to make profits. And that's a good thing. I like corporations to make profits. It means they employ more people, they invest in new research and development, they give to charitable causes, and billions of people around the world invest in their profitability, either directly or through their compulsory superannuation/pension fund. Some of them even pay taxes! There is nothing compelling corporations to ensure that their profit-making activities need to be ethical and advance societal and environmental benefits. Well, nothing except us. We are the investors, workers, social media trolls and voters who can make the world a better place. We are the checks and balances. And if we vote for leaders who deny the inconvenient truth of climate change, state that carbon dioxide is a weightless gas, or believe that building endless dams is the key to a bright future, then.... Sorry I went off task.

My real task in this chapter is to develop and thrash out the hypothesis that the major life science companies are, at the heart of things, chemical companies. Chemical companies who view the world in a chemical mindset. There is nothing intrinsically wrong with taking a chemical approach to understanding a lot of the things that happen in the bio-physical world. I hark back to Chapter 1 – almost everything that happens in our body is a complex set of chemical reactions. Without the beauty of redox reactions, we would have no biological energy in our cells.

Chemical companies are predominantly a child of the Industrial Revolution, when the manufacture of chemical compounds for steel, drugs, soaps, textiles, paints, dyes, fertilisers and foods moved from a cottage industry to one based in large high-throughput factories. As chemical processes became more and more advanced, synthetic compounds like celluloid, viscose and plastics began to be manufactured. By the end of World War One, some rather massive companies had formed, particularly in the USA (American Cyanamid, DuPont, Dow, and a little company called Monsanto, who mainly made chemicals used in the food industry), Germany (BASF, Agfa, Bayer, Höechst), Switzerland (CIBA, Sandoz and Geigy) and the UK (British Dyestuffs, Nobel Explosives, United Alkali). Many of these companies found that they could step up on the economies of scale (less costs more profit) by merging with one another. This was most notable in Germany with the creation of IG Farben in 1925 from a forced merger of Agfa, Bayer, BASF, Griesheim and Höechst). At the time, this industrial behemoth was not only the largest chemical company on earth, it was also the biggest company in Europe. Bearing Ramsay's words about "wealth and general prosperity" in mind, British companies merged together to form Imperial Chemical Industries (ICI) in 1927 from British Dyestuffs, Brunner, Mond and Co., Nobel Explosives and United Alkali). Perhaps they also had in mind that WWI had been deemed "The Chemists' War", or possibly that Alfred Nobel (a chemist) had said: "I intend to leave after my death, a large fund for the promotion of the peace idea, but I am sceptical as to its results". The following year a number of French companies combined to form Rhone-Poulenc.[1]

German chemist, Justus von Liebig, had first demonstrated that inorganic nutrients such as nitrogen, sulphur, phosphate and calcium were required for plant growth. Among many other things, Liebig is known as the "father of organic chemistry", and his work on fertilisers led to the understanding that crop yields could be significantly boosted by the addition of elements (NPK or nitrogen, phosphorus, potassium) in inorganic forms. For example, he became convinced that nitrogen in the form of ammonium could be taken up by plants, and wrote about it in his books on *Agricultural Chemistry* in the 1840s. By the mid-1800s, mining of various sources of these elements had become increasingly profitable, as farmers in Europe began to see the beneficial effects of these chemical fertilisers. Giessen University, where he worked in central Germany, was named Justus Liebig University in 1945 in his honour.

The first of these chemical fertilisers to have worldwide impact was superphosphate, developed by Englishman John Bennet Lawes. Lawes inherited a manor house known as Rothamsted from his father, and spent much of his life undertaking agricultural experimentation there. He developed superphosphate in 1842, which was the result of treating mined rock phosphate with sulphuric acid, rendering the phosphorus into a more plant-available form. The following year he established Rothamsted Experimental Station, where agricultural research still continues today with more than 500 staff. My university, the University of Queensland, has an agricultural campus at Lawes, named in his honour. This is where the Queensland Agricultural College was first founded in 1897, and has been part of the University of Queensland since 1989.

By and large, the chemical companies were predominantly focussed on household and industrial chemicals, and few had much interest in agriculture, with the exception of chemical fertilisers. Some were involved in conversion of the mined compounds into sources that seemed more suited to plant growth. Chemical research in Germany was going from strength to strength, with new breakthroughs being made every year, by creating totally novel compounds, or finding ways to chemically synthesise naturally occurring substances which were expensive and difficult to extract from their source, such as a mineral ore, or more commonly, a plant. The superphosphate experiments from Rothamsted were to have a major impact on agricultural productivity in Australia. The mining of guano from Pacific Islands like Nauru started in the late 19th century, and conversion to superphosphate enabled new improved pastures on the notoriously P-deficient Australian soils. While this had a positive impact on crop productivity, it was the wide availability of cheap superphosphate and subterranean clover that boosted the stocking capacity of many sheep properties by up to five-fold that increased sheep numbers. Australia was said to have been "riding on a sheep's back" for the first 50 years of the 20th century, when Australian merino wool was in its golden age. Stocking numbers had never been higher, peaking at over 180 million by 1970, driven by modern chemistry and plant breeding. However, modern synthetic fibres developed by chemical companies (such as nylon, acrylic, Terylene, polyester), many of which came out of DuPont factories, were soon to largely replace wool in the fashion stakes.

Speaking of clothing and fashion – can anybody honestly say they have never owned a pair of blue jeans? The blue colour most favoured for clothing, furniture and carpets was indigo, and the best source for this intense dye was the plant *Indigofera tinctoria*, domesticated and widely grown in India. India already had a lucrative indigo trade with Europe when the Greeks were the richest and most sophisticated society. So it won't surprise you to know that the name indigo derives from the word, *indikón,* the Greek word for India. This intense blue was highly sought after and was something of a luxury item. It was also extremely useful for dying cotton, hence the *coton de Nimes* or *serge de Nimes,* which has become one of the most widely worn types of cloth in the modern world, blue denim. By the end of the 19th century, over 7000 km^2 of arable land was used to grow indigo-producing plants. That's 7 million hectares! That's about the total area worldwide grown today for the production of bananas (the world's most eaten fruit) and tomatoes (the world's second most eaten vegetable after potato) COMBINED!

By 1880, German chemist Adolf von Baeyer had worked out two different ways to synthesise indigo, and by 1883 he finally worked out the chemical structure of the dye. This opened the way for mass production without tying up 7 million hectares of land (well, and also causing a massive change in agricultural land use in India). By 1897, a southern German company, Badische Anilin und Soda Fabrik, had worked out how to make indigo on a commercially viable scale. That is the company that now calls itself BASF "The Chemical Company" and by 2014 was the largest chemical company in the world, with sales of over $78 billion, almost $20 billion ahead of its biggest rival, Dow.

The real chemical breakthrough for agriculture and food production, however, came with the Haber–Bosch process, which enables the conversion of inert nitrogen gas (70% of the atmosphere) into ammonia. Ammonia is a gas that is then readily converted into urea. Urea became a cheaper, more concentrated source of nitrogen (just under 50% nitrogen), and led to much improved plant growth compared to "humus" or animal manures and composts. As a dry product, it was also much easier to transport around and spread over fields, easier to store, and just, well, convenience in a bag. Of course, ammonia may also be converted into compounds that are explosive, and during the first half of the 20th century, that was the main source of demand for the Haber–Bosch process. While the Green Revolution has been estimated to have saved around 300 million lives thanks to the efforts of Norman Borlaug, the biggest single contributor to the massive numbers of people on the planet is the Haber–Bosch process, which has been deemed to have saved over 2.7 billion lives (www.scienceheroes.com). Hence it was the chemical companies which drove much of the increase in agricultural productivity in the 20th century. Better genetics also played a major role, but for most crops, profitability just was not such an attractive prospect for investment. For example, you provide a farmer with a new variety of wheat, rice, potato or soybean. They grow the crop and if they like it, they save some of the seed (in the case of potatoes the tubers) to replant next year. They will buy the fertiliser from the chemical company every year, but not necessarily the seed. Only hybrid seed was an attractive money-spinner for seed companies, and hence most of the seed companies that were making money were producing locally adapted F1 hybrids of crops like maize, rapeseed, sorghum and sugar beet, which provided higher yield and resilience than the inbred lines.

Come the agricultural biotech revolution in the 1980s, most of the companies involved in agricultural productivity could be described as seed companies (like Pioneer, KWS, Advanta, DeKalb), machinery companies who made tractors, planters and harvesters (such as John Deere, Case, New Holland, Massey-Ferguson) or chemical companies who made many of the management tools in the form of insecticides, fungicides, herbicides or fertilisers (Monsanto, Ciba-Geigy, ICI, Bayer). And it was the chemical companies who moved the fastest to develop themselves into agbiotech companies. And in some ways that seems a little strange to me. Wouldn't it have been the seed companies who went down the biotech pathway as they were the companies packaging together novel and improved combinations of genes into seeds, cuttings and tubers? These small genetic packages were the means by which a farmer can access many of the improvements to productivity and quality. That would make total sense.

But what really made sense was the fact that the chemical companies were just so much bigger than the seed companies. They had the financial might, the infrastructure and the means to attract and retain the intellectual capital and the lawyers to protect their intellectual property. They had been doing it for decades, and they had a very long history of taking out patents (and defending them when the need arose) on new discoveries for things such as polymers and food additives, including household names like nylon, Teflon, Lycra, Kevlar, Aspartame

(NutraSweet) and AstroTurf, all of which people wore, drove on, sat on, played on or ate. Of this list, all of them were proprietary inventions of either DuPont or Monsanto.

At the beginning of the agbiotech revolution, there was not a single seed company anywhere near the world's largest companies. Way back in 1985, the top five companies in the world (according to the Fortune 500) were ExxonMobil, General Motors, Mobil, Ford and Texaco. Producing transport vehicles and the fuel that went into them were far and away the most effective ways to make monumental piles of money. One chemical company made it into the Top 10 – DuPont at number 7 with a reported revenue of $35.9 billion annually. Also making it into the top 50 were Dow ($11.4 billion), Union Carbide ($9.5 billion), with Monsanto and American Cyanamid making it into the top 100 with multi-billion dollar annual revenues. In 1985, the world's largest seed company was Pioneer, and they had an annual revenue of $177 million (according to the *New York Times*). By 1990, more chemical companies had made their way into the Fortune 100, particularly European-based corporations including Merck, Pfizer, Bayer, BASF and Lyondell.

In 1985, Pioneer Hi-Bred was the world's largest seed company and had over 35% of the most lucrative market in the world, the North American corn (maize) seed market. Historically their major rival was the even older company, DeKalb. This seed company was founded in 1912 in DeKalb, Illinois, as a grower cooperative. Like Pioneer, DeKalb had started to release hybrid corn seed in the 1930s, with yields of up to 20 bushels per acre (about 1.3 tonnes per hectare), better than the standard open-pollinated varieties. And just like Pioneer, DeKalb kept innovating with genetics, based on their forward-thinking decision to return a set percentage of profits into research and new product development. As a result, DeKalb produced the world's first hybrid sorghums, followed by hybrid sunflowers and alfalfa, and although they never surpassed Pioneer in the corn market, by the mid-1970s they had 23% of the corn market, and were getting close in size to Pioneer Hi-Bred to be an extremely serious rival as the world's largest seed company. But the world for seed companies was about to undergo revolutionary change. The biotechnology revolution led to serious disruption of the market, shortly followed by the genomics revolution, and it is still playing out today. The term "genetics" was coined in 1906, and within 100 years, most of the seed companies who started out harnessing genetics to make better crops no longer existed. If seed companies did survive, they had been taken over by chemical companies and hence the survival is in name only. Seed companies were just too small to survive as globalisation of the seed business became an unstoppable force.

To really mix the metaphors, and offend almost everyone, I'm now going to say that evolution gave way to revolution of biblical proportions. Any teenager who read the Old Testament was always struck by the large amount of "begatting" that went on back in the day. And not just the begatting but also the fact that many of these people were doing so when they were 800 years old. OK so the begatting moved a lot faster in the last 100 years than it did in Genesis. Obviously, these were more relaxed times back in the time of Genesis. And all those names and characters get so confusing that it's worse than trying to remember who's who in *War and Peace* or *Game of Thrones*. I can't go through all the various metamorphoses that seed and chemical companies have undergone in the last 100 years because:

1. There are too many iterations of all this begetting.
2. You would be bored and drift off to do something easier like watch Wagner's *The Ring Cycle*.
3. It's still going on so the information would be redundant or misleading or just wrong.

To illustrate the point, I will use personal experience. In Australia (like many other countries) we have a federal government Australian Research Council (ARC) who fund both basic and applied research. They have an excellent mechanism to get universities to work with industry (and industry to work with universities – something often overlooked) to help bring research into making an impact. The mechanism is called money. To be awarded an ARC linkage grant to undertake research, a university (or two) will partner with an industry partner (or two or more), and a percentage of the total funding must come from the industry partner. The first ever such project I led was a fantastic learning experience, to work on drought resistance mechanisms in canola with Australia's largest seed company,

Pacific Seeds. Pacific Seeds had been recently acquired by a larger entity, Advanta. But this was not a problem. In fact, it was a great advantage and selling point for the project because we had partners in Advanta Canada and Advanta Belgium on the project proposal. Being simple scientists who read the sports pages and not the finance pages, we were totally unprepared as to what would happen next. This all happened in 2005.

Monumental change was upon us. From the time we wrote and submitted the proposal to the time we were notified that our application was successful (about 8 months later), Advanta had been sold. Turns out Syngenta (a Swiss-based agbiotech company) had acquired control of the company by taking over the holding company Advanta B.V. in the Netherlands. They kept the North American corn and soybean business and sold the rest on to another party. It's a form of asset stripping – buy, take what you want, sell the rest for more than you paid for it. To make things more "interesting" the other party was not a chemical company with a clear plan for future activities. They had been sold to a corporate raider, an equity fund known as Fox Paine. Or "Foxing Paine in the ...." as they rapidly became known to us. It's not fair to say they didn't have a plan. They had a plan to make shiploads of beautiful dollars by dividing up a company and selling the parts for more than the whole. Way more.

I'm happy to stand in front of a class of students and explain how *Agrobacterium* works, or how wheat evolved in farmers' fields. But to pretend I understand the way international finance works would be like claiming the world is 6300 years old because I read it in a book, and applied some simple maths to all the begatting that went on. Anyway, from what I can read from various sources, not everything was sold because Advanta Lambda B.V. retained some of the businesses, and transferred their shareholding back to Advanta Netherlands Holdings in 2005. You can look up what a holding company is. It is a company that does not produce goods or services. I think that means they are primarily about dodging tax, but I may be over-simplifying. There are lots of holding companies registered in places like the Cayman Islands, Luxembourg, the Isle of Man and Switzerland. Some may not be holding much, just a few simple things like large yachts, villas on the Cote d'Azur, or fleets of Ferraris and Porsches.

Enter a new player in the holding company stakes. An Indian agrochemical company known as United Phosphates Limited (UPL) (India). In 2006, UPL had a holding company called Bio-Win Corporation based in Mauritius (well why ever not?), who acquired Advanta Netherlands B.V. from Fox Paine. Advanta India sold their shares to Bio-Win, who then became sole owner of Advanta India. Advanta India then bought Advanta Netherlands B.V. who then bought Advanta Netherlands Holdings from Bio-Win. They (I can't remember who "they" is by now with all these holding companies) then took over Advanta Netherlands B.V., Advanta Finance B.V., Advanta International B.V., Pacific Seeds (Thailand), Advanta Semillas (Argentina), Pacific Seeds (Australia) and Advanta Africa Seeds. Then UPL acquired the shareholding of Bio-Win in Advanta India, and by the end of all that UPL then owned all the companies. I think . . . . And Bio-Win, the holding company, eventually changed its name to UPL Corporation Limited and is a wholly owned subsidiary of UPL Limited. I think that means it's still a holding company but may not actually be holding anything at the moment. And if you are not confused by all of this, I suspect you must be holding a few holding companies yourself, so let's just leave it at that.

So, once our project with Pacific Seeds started, we were deprived of two very important partners. Advanta Canada was sold off to Monsanto, and they were no longer able to collaborate. Advanta in Europe was carved up and the canola programme was sold to a French company, Groupe Limagrain. They were going to perform the expensive DNA marker work for us as an in-kind contribution to the programme, but could no longer do so. Needless to say, despite our best efforts the project did not get as far as we hoped without the genetic and lab resources originally envisaged.

However, we developed a good working relationship with the company and in 2007 we negotiated another project to improve sorghum grain quality and forage for biomass, which started in 2008. By the time the project finished 4 years later, the grain sorghum breeder had retired, the forage sorghum breeder was retrenched, then the replacement grain sorghum breeder left in disgust and went to St Louis to work for Monsanto. The Global R&D Manager, a wonderful man based in Toowoomba who instigated most of the project, retired and became an avocado farmer. A new Global R&D Manager was appointed, who after a short time moved from Argentina to Dubai, where presumably he works for a holding company? The forage sorghum breeding activities were moved to the Punjab in northern India, and when we tried to negotiate a follow-up project we were told that Advanta were going to continue the research in their bio-technology centre in Argentina because it was cheaper. In early 2017 I heard from the Advanta people in Argentina that the Balcarce biotechnology centre was to be closed and all activities would move to the USA. Just a small personal anecdote on the life and times of the current agbiotech scene, and these sorts of things continue to happen right now. The seed and agro-chemical companies who undertake most of the agbiotech research in the private sector continue to invest and divest. Sometimes they get others like Fox Paine to help them. According to their website, Fox Paine "specializes in providing solutions and capital for management buyouts, public-to-private transactions and growth capital initiatives. The firm participates in friendly, co-operative transactions with the goal of creating tangible benefits for all parties involved". Everyone's a winner baby ... except the poor people who just do the work!

The top 10 biggest charts for these companies are in a con-stant state of flux, but for illustrative purposes I have shown figures from a few years ago (Table 4.1). Philip Howard from Michigan State University wrote a great piece in 2009 attempting to visualise the takeovers in the agbiotech agrochemicals space (Figure 4.1).[2] Although fantastic, his figures are already seriously outdated, but it is good to reproduce here. What is evident is that some players have been significantly more acquisitive than others. You don't have to look at the figures too critically to reach the conclusions:

1. BASF, the world's biggest chemical company, has not been very active in acquiring others.
2. Monsanto, the world's biggest seed company, may well have got there by acquiring as many other companies as possible.

Monsanto, who we have already established were there right at the start of the agbiotech revolution, moved fairly quickly to get into the biggest markets. They managed to get some pretty good

Table 4.1 Largest chemical, agrochemical and agricultural seeds companies.

Chemicals (2014)	Agrochemicals (2013)	Seeds (2013)
1. BASF	1. Syngenta	1. Monsanto
2. Dow	2. Bayer Crop Sci	2. DuPont Pioneer
3. Sinopec	3. BASF	3. Syngenta
4. SABIC	4. Dow Agro	4. Groupe Limagrain
5. Exxon Mobil	5. Monsanto	5. KWS
6. Formosa Plastics	6. DuPont	6. Land O Lakes
7. Lyondell Basell	7. Adama	7. Bayer Crop Sci
8. DuPont	8. NuFarm	8. Dow Agro
9. Ineous	9. FMC	9. Sakata
10. Bayer	10. UPL	10. DLF-Trifolium

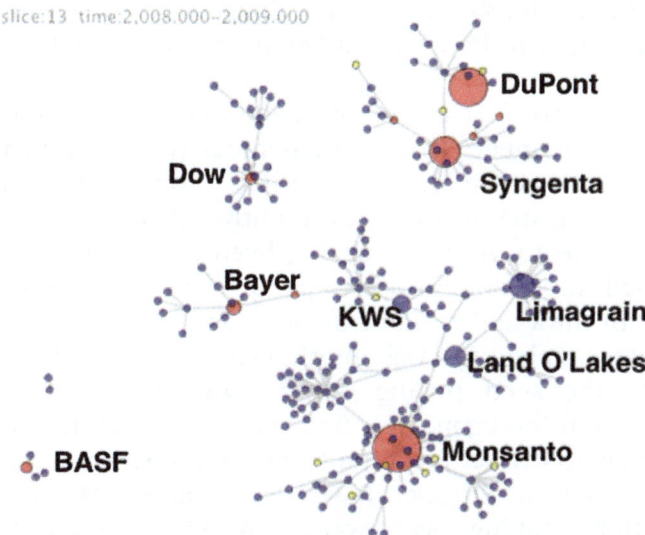

Figure 4.1 Largest chemical, agrochemical and agricultural seeds companies are increasingly networked through agreements to cross-licence transgenic seed traits.
Reproduced from ref. 2 with permission from Philip H. Howard.

technologies by acquiring one of the other pioneers, Calgene, who developed the first commercial GM crop plant, the FlavrSavr tomato. Its major step into the corn market was the acquisition of DeKalb in 1995. And to ensure they had a big toe in the water when it came to cotton, they took over Agracetus, Delta and Pine Land, and Mahyco (India). Monsanto got into the canola market with the acquisition of Advanta North America.

Similarly, one of the "new kids on the block", and the biggest agrochemical company at the time, Swiss giant Syngenta, made it big by effectively acquiring AstraZeneca (formerly ICI Seeds), Zeneca, Sandoz, Novartis and Ciba-Geigy. Syngenta acquired corn and soy with the part of Advanta North America that used to be Garst. Garst was another one of those companies like Pioneer and DeKalb, who started as a little company in 1931 in Coon Rapids, Iowa, to make hybrid corn. In 2014 the Garst name was "retired" by Syngenta and was replaced by Golden Harvest, where they now have about 9% of the North American corn seed market. Presumably, like Michael Jordan's #23 Chicago Bulls jersey hanging up on many bedroom and boardroom walls around the world, the Garst name lingers on as a rusty old metal advertisement on the wall of a few dilapidated sheds across the Midwest.

DuPont decided to just go for it, and went after the biggest fish of them all, Pioneer Hi-Bred. Remember their relative revenues in the mid-1980s (DuPont $35.9 billion, Pioneer $117 million). It was akin to a massive orca accidently getting a small herring stuck in its teeth. And the herring lived on. The two companies continued to operate their separate agbiotech programmes, Pioneer in Iowa and DuPont in Delaware.

Another US chemical behemoth, Dow, decided they had better get into the seed business and acquired Sudwest Saat in Germany and Mycogen. Dow had started in 1897 in Michigan as a company to extract bromine from groundwater.

Meanwhile in Europe, Bayer was leading the charge for acquisitions, taking over Aventis, AgrEvo and Plant Genetic Systems, the company started in Belgium by Marc van Montagu, the GM plant pioneer. Well actually, Plant Genetic Systems was taken over in 1996 by AgrEvo, who then merged with Rhone-Poulenc in 2000 to form Aventis, who were then acquired by Bayer in 2002 to form Bayer Crop Science.

And it has all changed by 2017/18, when I am writing this. By 2010, the "big six" agbiotech-seed-chem companies were Monsanto, DuPont Pioneer, Dow, Syngenta, BASF and Bayer. Many of the smaller agbiotech and seed companies had been acquired by these six, and most had ceased to continue as they were being captured for their genetic materials, their biotech IP, or both. Suddenly there weren't too many tasty herrings left, and it seemed that the big fish were starting to circle one another. The gloves were off, which is a difficult thing when you're a fish, or an orca. OK, I have mixed my metaphors there, but it's water off a duck's back. Or as one of my collaborators from India always used to say with a dismissive and amiable head wobble "it's duck's water".

Prices for agricultural products (cereals, oilseeds, fruits, vegetables) generally continue to follow the downward price trend in real terms they have for much of the past few decades. With the exception of perturbations such as major droughts, hurricanes, cyclones and floods, which predominantly have their impact at a local or regional level, prices for most products (even avocadoes) do not usually have major increases. And they certainly go down as a percentage of disposable household incomes as consumers become more affluent. The FAO price index is calculated as an average of five major commodities (meat, dairy, cereals, vegetable oils and sugar). The index was set at 100 in 2002–2004, and showed that back in 1962, the index was 129. There was a fairly significant spike in 1974 when the index reached 177. For much of the 1980s to 2006, the index hovered in the 90–110 bracket, before starting to rise and peak at 169 in 2011. It is now back to 129 in 2017, exactly where it was in 1962. One of the biggest drivers of this is the cost of cereals. They are not only the major staple for human consumption, but when prices go up, the cost of producing meat and dairy also goes up. This is most noticeable in the price of chicken and pork, because these animals do not have "grass-fed" as an option.

An article in *Forbes* (3 February 2012) pointed out that as food prices start to fall, farmers become more careful about their input costs, especially fertiliser and pesticides. If crop prices are lower than last year, the farmer may not put on quite as much fertiliser as last year, and she may choose to ignore the insects and fungus on her crop and settle for a lower yield that year. This then squeezes the producers of these inputs, and they start to look for ways to become more efficient, one of which is upping the economies of scale.

So it came to pass that Monsanto, the acquisition king in agriculture, decided to make a bid for Syngenta in 2015. In May of the same year the offer of $45 billion was rejected. However, the blue touch paper had been lit, and it started such a ferocious frenzy of takeovers, mergers and alliances that one day some-body will make a movie out of it. After the first exciting episode, where nothing of real consequence happened, Dow and DuPont announced a merger in late 2015. This $130 billion merger was finally completed in September 2017. Not to be perturbed, a hitherto little-known player, ChemChina (actually a state-owned corporation, China National Chemical Corporation) then made a bid for Syngenta, upping the failed Monsanto bid to $46 billion, which Syngenta accepted. The acquisition was finalised in May 2017, and no sooner had the ink dried that it was announced that ChemChina and another state-owned enterprise, SinoChem, would have a go at merging. Such a merger would make a company worth around $120 billion, and suddenly BASF was going to be only the second largest chemical company.

The DowDuPontPioneer megalith recently opted to spin off its agbiotech arm, leading to a totally new seed and agrichemical company, Corteva Agriscience. The new Corteva encompasses the old Pioneer seed business, DuPont Crop Protection and Dow AgriSciences. Apparently Corteva is a combination of words meaning "heart" and "nature".

But wait, there's more. Monsanto was still licking its wounds after its inability to clinch the Syngenta deal when it got an offer. It was from Bayer, who proposed to spend $66 billion to take over Monsanto. Monsanto took a few deep breaths, and finally in September 2016 it accepted the offer. Between Bayer and Monsanto, they control 25% of the agrochem-seed market, and a considerable swathe of the agbiotech patents.

Now, as we would like to hope, there is significant government oversight of these sorts of megadeals. There can be a real danger of the development of a monopoly. To use a hypothetical ex-ample, let's imagine a situation wherein something everyone relies on has the potential to be owned by just one company. It could be something quite humble and unsexy. For the sake of illustration, imagine there were only two companies who made cardboard boxes. Yep, cardboard boxes. These are something that the average consumer, unless they are moving house, never thinks much about, and never actually purchases. For most of our workplaces they are something that are of fleeting annoy-ance. They are bulky, can be hard to open, and in the home and workplace cardboard boxes become a bit of a nuisance problem.

Except again when you move house. How many scientists out there have moved house with surplus boxes from work – you move house with the labelled cardboard boxes from Bio-Rad, Thermo Fischer and Illumina. Or if you work in a bar, with boxes labelled with Coca-Cola, San Pellegrino and Heineken. Oh, and those boxes from Amazon because, after all, who shops in the local retail centre anymore. Then many of these boxes sit, unopened, in the garage or basement. This is all absolutely fine because conveniently, when you need to move again 3 years later, you don't even have to unpack them. You just have to package up the guilt associated with "having too much stuff" and being "too busy to declutter". Heck, some of the boxes may even contain the Southern blot and sequencing gel autorads from your research back in 1987, but you just can't bring yourself to throw them out. This is getting way too personal, so back to those hypothetical cardboard boxes.

While to you they are not thought about much, they are an increasing part of world commerce, and quite a profitable little business they represent. For those industries who need to actually buy them, they are part of the price structure. The more expensive they are, the more the price needs to be factored into what you pay for that pair of shoes, gorgeous silk blouse, or artificially aged wooden sign inscribed with "Dance like nobody is watching" or "HOME" (just in case you forget where you are). Competition is necessary to drive down prices – or that's what our politicians tell us. Economic theory tells us this should be the case. Reality is somewhat patchy at times when it comes to real world evidence (health insurance, electricity, *etc.*).

In the case where you don't have a monopoly, competition can sometimes be fierce, possibly even volatile or unpredictable. This creates all sorts of uncertainties, and can wreak havoc with sales and profit margins. So it came to pass in Australia with the two largest cardboard box manufacturers, Visy and Amcor. These two companies between them had cornered around 90% of the Australian cardboard box market, worth around $2 billion annually. From 2000 to 2004, the two companies colluded to quote the same price to effectively create a cartel, which is, in reality, a *de facto* monopoly. Senior executives from both companies would meet in parks or hotels, and contact one another *via* pre-paid phones. The major activity of this cartel was to ensure that whenever a major customer wanted to renegotiate the price, for example with Visy, they would approach Amcor for a price. The two companies would secretly share information and hence Amcor would quote a higher price than the existing price the company was getting from Visy. This cartel activity is known as cover pricing.

After it had been going on for almost 5 years, it was accidently exposed and Amcor admitted to the Australian regulators that they had a price-fixing deal with Visy – only after Amcor negotiated immunity from prosecution. It took Visy some time to admit that they were, indeed, colluding with Amcor, and in the end they were fined $36 million, and some of the senior executives were forced to pay a total of $2 million in fines. The owner of Visy, Richard Pratt, was one of Australia's richest men, with a personal fortune of $5.5 billion when he died in 2009. It has been estimated that his company cheated customers out of over $700 million (reported in the Herald Sun in August 2007) during the cartel's successful clandestine operation. As I read it, a loss (investment) of $36 million for a $700 million return is like backing the winner of the Kentucky Derby at 20 to 1 odds. After Richard Pratt died, his son Anthony Pratt took over the business. In less than 10 years he has managed to more than double the family wealth, and in 2017 he topped Australia's wealthiest list with a mere $12.6 billion. Obviously still backing lots of winners.

I hope this makes you look at the next cardboard box you see with a lot more respect! One could also reach the conclusion that crime does most certainly pay, even if you're caught.

Both US and EU regulators were concerned that a Bayer–Monsanto conglomerate would have way too much of the seed, trait and agrochemical business for the merger to be accepted. For example, the two main herbicide resistance traits were Roundup Ready (glyphosate resistance owned by Monsanto) and Liberty Link (glufosinate resistance owned by Bayer). For one large company to own the two most successful and most widely grown herbicide resistance traits on the planet could create a worrisome situation. Given that 88% of the GM crop area of 185 million hectares is grown to herbicide-resistant crops or so-called "stacked traits" (herbicide + insect resistant) crops, herbicide resistance traits represent the biggest market there is when it comes to plant biotechnology. Now that is a market that any seed/chemical company would drool over! You can imagine the government regulators looking at that market, then looking at who owned the traits, and reaching a well-founded conclusion that a Bayer–Monsanto conglomerate may have just a little too much of the action to ensure there would be sufficient competition in the market.

EU regulators cleared the Bayer takeover of Monsanto in March 2018, subject to a number of conditions. Bayer had to sell its seeds business in its entirety. BASF, who did not have a particularly strong holding in the seeds area, snapped it up for a cool €6 billion, with most of the business based on oilseed rape (canola), wheat, cotton and soybean. Bayer also had to divest itself of its vegetable seeds business, its pesticides and their traits. In addition, the EU insisted that Bayer sell its "research activities to develop a challenger product to Monsanto's glyphosate". Given that glyphosate is no longer on patent, and hence not really Monsanto's any more, this is an interesting one. However, I suppose Monsanto already has a fairly large project to develop the next-generation glyphosate, so it was deemed that they could not combine that with Bayer's programme. All of these rulings have knock-on effects with activities around the globe. Australian grain growers pay levies when they deliver their harvest to the local grain dealers. They use this levy to fund further research through a leveraged system with the federal government, leading to a rural R&D corporation known as the Grains Research and Development Corporation (GRDC). Now I'm not going to say anything bad about GRDC because it funds my research. No, that's not true. It invests in my research and partners with me – that's how they prefer to see it, and it is indeed true. However, in 2015, it also made a big partnership and investment with Bayer CropScience. They signed a 5-year agreement to establish the Herbicide Innovation Partnership, with an investment of $45 million. It's not entirely certain at present what the future of the Partnership will be.

What will be the fate of Monsanto? Will the name totally disappear sometime in 2018? This would be a terrible body blow for the anti-GM activists. What would they do with their campaigns like "Millions Against Monsanto"? I guess they could straight away up the fake news ante and go straight for "Billions Against Bayer". Just as well the Danish seed company Trifolium didn't take over Monsanto. "Trillions against Trifolium" would require significant recruitment of activists from other solar systems to make up the numbers. As you are no doubt aware, Monsanto is the "world's most evil corporation". Just do a quick Google (also on most lists of evil corporations) search and you can find out all about their wicked history.

One of my favourites, a site called Collective Evolution, tells us that this evil corporation is responsible for Agent Orange, DDT, asbestos and aspartame. That's a formidable list! And it does indeed reek of evil, except for a few facts that get in the way.

Fact 1: Agent Orange consists of a mixture of two chemicals, 2,4-D and 2,4,5-T. Both of these are synthetic auxins and were first developed by the US military in the 1940s. Neither is regarded as having high toxicity or carcinogenic effects on humans. In fact, 2,4-D is widely used as a selective herbicide, given that it will kill broadleaf weeds at a significantly lower concentration than it will kill grasses. Hence its use on lawns, golf courses and cereal crops is common. As a result of its higher toxicity and implications in being carcinogenic, 2,4,5-T was banned from use on food crops in the 1970s and was effectively banned in 1985 in most parts of the world. However, during the manufacture of these chemicals, a small amount of dioxin is made. While 2,4-D and 2,4,5-T break down under natural conditions, the dioxin does not. It is quite stable and will actually accumulate in things such as food, and after ingestion will accumulate in fatty tissues. From there, dioxins can cause reproductive and developmental problems, and may lead to cancers. What role did Monsanto play? It was one of more than a dozen corporations who made Agent Orange for the US military during the 1960s. Others such as Dow, Hercules and Uniroyal also made Agent Orange and a number of other defoliants for the US military to use in Vietnam. It should also be pointed out that the use of these defoliants was pioneered by the British, first to test the efficacy against reducing tsetse flies in Kenya and Tanzania in 1952/53, then later to test the efficacy against reducing hiding places for jungle guerrilla fighters in Malaysia during the Malay Emergency in the 1950s.

Fact 2: DDT, the insecticide that Rachel Carson exposed as an ecological disaster in the making in *Silent Spring* in 1962 was already almost 100 years old by then. However, it wasn't until 1939 that a Swiss chemist, Paul Müller, discovered that this chemical had insecticidal activity, and then it became widely used to stop the spread of malaria in the tropics during World War Two by both military and civilian populations. As Carson highlighted in her book, it was implicated in eggshell thinning and hence could lead to the eventual extinction of bird populations, leading to a silent spring. From 1950 to 1980 it was manufactured by 15 different US chemical companies, one of whom was Monsanto. So without this evil corporation, there were still 14 companies making DDT.

Fact 3: Asbestos is a naturally occurring mineral. It has been mined for over 4000 years. It was first described by Pliny the Elder, who died in the year 79. Asbestos is a mineral, in fact a family of six different fibrous silicate minerals, all of which are now known to be carcinogenic. However, before they were known to be dangerous to human health, they were widely used in building materials, especially given the superior qualities of asbestos-containing materials in retarding fire and having excellent heat insulation. What did Monsanto have to do with that? Your guess is as good as mine.

Fact 4: Aspartame is an artificial sweetener, based on the chemical modification of the amino acid phenylalanine. It was developed by the British company, G.D. Searle, in 1965. That company was acquired by Monsanto in 1985, 6 years before the aspartame patent expired and they set up a spin-off called NutraSweet to market the compound for food uses. Over 100 different food regulatory agents have deemed the product to be safe for food use, and many studies have failed to find any link to adverse health outcomes, including cancer, neurological and psychological disorders and problems with breastfeeding. Monsanto sold the company on in 2000 and today the largest producer of aspartame (under the new label AminoSweet) is Ajinimoto, a Japanese food chemical company.

So, Collective Evolution have really nailed it here. What have they nailed? They have nailed the fact that:

- they didn't want the facts to get in the way of a good story, or
- they don't care about facts.

One of their more recent achievements is the production of a film called *Sacrificial Virgins* which is about "young girls being severely damaged by the HPV Gardasil vaccine". After all, I'm sure cervical cancer is a much more preferable outcome than having a needle. This one really gets me for two reasons. The main developer of Gardasil is Ian Frazer, a Professor at the University of Queensland, and one of the nicest people you could ever hope to meet. Further to that, his own Frazer Family Foundation (set up to collect the revenue from his royalty as an inventor) has paid for the roll-out of the vaccination to the entire female population of Vanuatu. In partnership with UQ, WHO and the Gates Foundation, he has foregone the royalty for the provision of Gardasil to developing countries and ensured that the cost of the vaccination is held at $5. I really don't understand the way some people tick, but Collective Evolution are, to my mind, evil.

I spoke to Cami Ryan, the Social Science Lead at Monsanto. Cami, originally an academic from Saskatchewan in Canada, channelled Vivian (played by Julia Roberts with that smile) from *Pretty Woman*: "the bad stuff is easier to believe, you ever notice that?" Cami started her working life in a small biotech company in Saskatoon, a company cloning fruit trees and medical marijuana. She was encouraged to "go back to school" and one thing led to another and she found herself doing a PhD in social sciences. Her main interest at first was how do social relationships among scientists result in productivity and innovation? This then led her to look at the anti-biotech activists, and she was able to use her skills to show that the non-scientist activist networks were much more adept at organising and tapping into how the public can then be co-opted to "fill in the information holes in their own mind". Cami told me that as a species, humans are risk-averse and resist change. We particularly resist technological change, especially if at first glance we cannot see a personal benefit.

She believes that the Chemical Hearts of the powers within the biotech companies in the 1990s had their approach all wrong, and hugely underestimated the role that anti-GM activists would play. The biotech and food companies, if they had their time again, rather than resist the labelling of GM food, would embrace it. Rather than taking the approach of "it doesn't really matter" or that labelling would be too costly, the technology may have been better received if it was originally marketed with a positive label to demonstrate the environmental and food safety benefits. Almost 30 years later, as you will see in the following chapters, we have many measures of the food safety and environmentally beneficial outcomes of GM crops and foods. Unfortunately, in many people's minds, the proverbial horse may well have bolted!

BIBLIOGRAPHY

Richard Pilcher, Secretary and Registrar of the British Institute of
Chemistry in 1917.

REFERENCES

1. M. Forster, Chemistry in Modern Warfare, *Curr. Sci.*, 1936, **4**, 536–538.
2. P. H. Howard, *Sustainability*, 2009, **1**, 1266–1287.

CHAPTER 5

Wide Open Spaces

"Agriculture is the most healthful, most useful, and most noble employment of man."

Thomas Jefferson

"Every one of us that's not a farmer, is not a farmer because we have farmers. We delegate the responsibility of feeding our families to a relatively small percentage of this country..."

Tom Vilsack, US Secretary of Agriculture[1]

You have already seen the figures on the impact of GM crops on agriculture in Chapter 3. Figures are one thing. What about the experience of the people who grow the stuff to make a living? Are they happier, richer, wiser, more fulfilled because they are growing GM crops? Maria von Trapp (Julie Andrews) famously sang: "Let's start at the very beginning, a very good place to start..." Ignoring that advice, I'm going to start in perhaps the worst possible place. A place where GM crops have been effectively banned and, according to Matin Qaim (see below), where "80% of the population is anti-GM and the rest don't care". Germany.

In the historic central German city of Göttingen (Figure 5.1) is the just as historic Georg-August-Göttingen University, named after its founder, King George II of England, which makes

Good Enough to Eat? Next Generation GM Crops
By Ian D. Godwin
© Ian D. Godwin 2019
Published by the Royal Society of Chemistry, www.rsc.org

Figure 5.1   Göttingen viewed from the top of the city's cathedral.

total sense. Well it starts to make more sense when you realise he was also the Elector of Hanover, and in fact, a German. The University has always been one of Germany's strongest scientific universities, never more so than between 1920 and 1933 when it was a world-leading institution for theoretical physics, attracting staff and students from all over the world, including J. Robert Oppenheimer, who went on to lead the Manhattan Project to make and use the first atomic bomb. This uber-physics dominance came to an abrupt end when a certain A. Hitler clamped down on theoretical physics, or what he termed "Jewish physics", and purged the University of many of its brightest scientific minds. Many of them went on to become the brightest minds in universities and research institutes in the UK and USA. Jewish physics was one of Germany's highest impact exports, and Göttingen provided three Nobel Laureates in Physics to their new countries, James Franck (1925), Max Born (1954) and Eugene Wigner (1963). Historic Göttingen was also the professional home of the great spinners of fairy tales, the Brothers Grimm, and the Iron Chancellor of the new united Germany, Otto von Bismarck.

Matin Qaim is Professor of International Food Economics and Rural Development at Göttingen University. He is an outspoken advocate for the responsible use of biotechnology for food security and regional development and among many other activities, a member of the Humanitarian Board for the Golden Rice Project. He's not a spinner of fairy tales at all – he deals in serious figures, and large-scale, long-term analyses of how the growth of GM crops has impacted on the livelihoods and markets of resource-poor farmers in Africa, Latin America and Asia. Matin was driven by international development issues and particularly the role new technologies could play in improving the livelihoods of smallholder farmers living in poverty. While studying for his PhD at Hohenheim, he was encouraged to look at the "new technology" of GM crops, and what impact (positive or negative) it may have on farmers in developing countries. As a postdoc at University of California Berkeley, he started a long-term analysis of Bt cotton and smallholder farmers in India. His study group involved a total of 64 villages in four cotton-growing provinces of India.

Bt cotton was first commercialised in 2002 in India, although "pirate" or "illegal" Bt cottons were already being grown. The first release introduced three hybrids with Bollgard I technology, from Monsanto and their Indian partner Mahyco (soon to become a wholly owned Monsanto subsidiary). How did this play out? By 2011, over 95% of all cotton grown in India was Bt cotton, and over 880 varieties and hybrids were available to farmers *legally*. There was always an important and difficult-to-measure black market of varieties too, some of which were not Bt cottons at all and generally led to some fairly devastating crop losses. Within a few short years, the outcomes of Bt cotton were plain to see. For example in Kanzara, a village in Maharashtra, there were 305 households, 203 of which had land for agricultural production. Average farm size was a mere 4.7 acres (just over 2 hectares). In the first year, only eight farmers adopted Bt cotton. Table 5.1 shows the impact, and the reasons why most farmers had adopted Bt cotton three seasons later.

So within the first year, yield was up (34%), insecticide use was halved, and as a result net income was up 70%. By 2006, yield had almost doubled, insecticide use was down to 30%, and net revenues were more than doubled. This was a remarkable achievement, and for cotton farmers was the green revolution they missed out on last century. Even when averaged across all India, by 2008 cotton yields were up by 24% and profits were up

Table 5.1 Impact of the cultivation of Bt cotton on key parameters in Kanzara (modified after Subramanian and Qaim, 2010).[2]

| | Season 2002/03 | | Season 2006/07 | |
	Bt cotton	Conventional	Bt cotton	Conventional
Insecticide use (t per acre)	2.07	4.17	1.22	1.55
Yield (kg per acre)	658	491	842	590
Net revenue (Rs per acre)	5294	3132	7120	4181

by 50%.[3] This meant households had more cash for consumption for more diverse foods, shoes, clothing and education. For the most vulnerable farmers, household incomes had increased by over 130%.[2] And for those of you who enjoy data, you will also have noticed that because most farmers had adopted Bt cotton, even those growing conventional cotton enjoyed much better yields and had to spray less. The insect populations were lower because most farmers were growing Bt cotton, so the Bt cottons had a prophylactic role for the whole region, although farmers growing conventional cotton were still almost Rp 3000 per acre behind the Bt adopters in net revenue terms. For an average-sized farm, this meant the Bt adopters were ahead by almost Rp 14 000 (US$320 in 2006).

According to K. R. Kranthi, the Director of the Central Institute of Cotton Research (CICR), it should have been even better. Monsanto's original Bt cotton lines were not ideal for many Indian farmers. The first Bollgard cottons were in long duration varieties. In many crops, provided water is non-limiting (a big assumption when the crop is not irrigated), a longer duration will often mean a higher yield. A downside to this was that the longer growing season meant more exposure to insects, including other insects not targeted by the Bt toxin. In 2015, growth conditions were perfect for whitefly, which destroyed over half the cotton grown in Punjab and Haryana.

And it hasn't been the easiest situation for the seed companies either. The rivers of gold that they expected in royalties were not quite as expected, because of government regulations (interference as they would say), the availability of publicly funded CICR varieties at much lower cost, and the readily available illegal Bt cottons. As widely reported, many of the cotton seed companies believe they have been getting a raw deal. It is also a fact that resistance management has been much more challenging in India compared to jurisdictions like Australia or the USA. Although farmers were compelled, and usually contractually obliged, to grow refugia, this was often not done, or if it was done, farmers would spray the refugia with insecticide as it was hard to watch all that productivity get eaten for a seemingly academic goal of slowing down the development of Bt resistance in the insect. In addition, Indian farmers still faced fierce international competition. Although India has the largest area of cotton under cultivation of any country, competitors like China continued to produce more cotton on less than half the area of land. However, for farmers, there was a general sense that Bt cotton was a technological great leap forward.

This also appeared to be the case for Africa's largest cotton-producing country, Burkina Faso. Like India, most producers are resource-poor smallholders. Cotton, however, provides jobs for 17% of the workforce and is Burkina Faso's most important export after gold. Cotton is so important it is known locally as white gold. And as history shows, wherever there is a gold rush, new people are attracted to take part from all over the world. Monsanto was obviously attracted to become an integral part of the gold rush. Monsanto entered the market in Burkina Faso, and in striking similarity to India, made the first commercial release of three cultivars in 2008. The new cultivars were variants of locally adapted lines using a conventional technique known as backcrossing to introgress the Bt gene. The theory behind backcrossing is that you end up with a cultivar that is indistinguishable from the original elite line, except for the fact that it has one new trait. In this instance the trait was insect resistance courtesy of the Bt gene. Within 5 years, over 70% of all cotton was a Bt line. For the average farmer adopting the new technology on their 3 hectare farm, they achieved a 22% increase in yield and a 51% increase in profit.[4] Farmers loved the product and cotton production in Burkina Faso became the poster child for the positive impacts of GM technology. However, not everyone was happy.... In fact, the people buying the cotton from farmers were positively displeased – so displeased that they lobbied the government, finally succeeding in getting it to agree to phase out Bt cotton and return to conventional cotton by the 2017/18 season. Where did it all go wrong?

The first thing that is required is an understanding of the process of backcrossing. Backcrossing is a commonly used plant breeding technique to make a small incremental improvement to an already elite cultivar. So if you have an elite cotton line with good yield, disease resistance, cotton quality, and good local adaptation (to temperature, soil, rainfall, *etc.*), "if it ain't broke, don't fix it"? The answer to that is if you have a single trait, like Bt insect resistance, but that trait is in a cultivar well adapted to Texas, like DP 555, you can use the US line as a donor of the Bt gene and the elite locally adapted line, say FK290. So you cross FK290 with DP 555, and the resulting plant has 50% of its DNA from FK290 and 50% from DP 555. You then "backcross" this to FK290 and it has, on average, 75% of its DNA from FK290 (the recurrent parent). The more you backcross, the more the plant actually starts to resemble the recurrent parent. Or put another way, with every successive generation, the backcross progenies will be 87.5%, 93.75%, 96.875%, 98.5%, 99.25% like the recurrent parent and so on, and all you need to do is make sure they have the Bt gene. Hence for successful backcrossing it is usual practice to backcross for more than six generations so that the DNA from the donor parent is less than 1% of the genome.

As does happen in gold rushes, the Monsanto cotton breeders were in a hurry, and they only carried out three generations of backcrossing before the new Bt lines were released. So our figures would say that on average these new Bollgard lines were still only 93.75% like the elite parent (let's use FK290 as the example again). But they seemed to be pretty good, and farmers certainly embraced them. When as a farmer you buy a variety that you have been told is similar to FK290 but is insect resistant, you know exactly how to manage it as a crop, and what it should yield and pay. So far, so good. Within 5 years, over 70% of all cotton in Burkina Faso was Bt cotton. The average smallholder farmer on 3 hectares enjoyed a 22% increase in yield and a 51% gain in profits. Everybody was happy.

Wrong, some people were unhappy. And not just Greenpeace. Once the cotton left the farm, a few problems started to occur. Cotton from West Africa is regarded as some of the best in the world, and cotton from Burkina Faso was the *crème de la crème*, attracting a hefty premium price. Their cotton was highly prized because it was hand-picked and, as a major bonus, decades of breeding in national programmes ensured that the fibres (known as the staple) were long, strong and uniform. In the new Bt cultivars, the cotton was strong, it was uniform but it was no longer long. In fact, the Bt cottons had a staple length of 1/32 of an inch shorter than the elite parents. The Monsanto breeders regarded this as non-consequential. After all, what's 0.8 mm between friends?

Turns out, as Maxwell Smart would say, the Bt cottons "missed it by that much", even if it was only 0.8 mm. This tiny little measurement (about half the size of a pinhead) was enough to lose Burkina Faso's market at the high end of the quality spectrum. So even though the farmers remain very positive about the great on-farm benefits, the cotton marketers were up in arms, losing their lucrative international contracts, and creating a glut of unsold cotton. By 2015, Burkina Faso was producing 700 000 tonnes of cotton, almost double pre-GM production. Yet the marketers were finding their new product difficult to sell to their traditional markets because the quality was not what it once was. The Burkina Faso government responded to this quickly, because after all this was an export market worth $400 million, and if it could not be sold easily, many people would be out of work. In a country where average annual household income was around $790, $400 million can go a long way. The most rapid resolution was to rule that GM cotton be phased out and replaced by conventional cotton by 2017/18.

So in the rush to get a product to market, shortcuts were taken. By taking an extra year of backcrossing, the GM cotton lines would have been indistinguishable (or at least 98.5% similar) from the original elite lines. But the reputation of Bt cotton was destroyed in Burkina Faso, and may not return for a long time, if ever. Most of the textbooks on plant breeding say the top three breeding objectives are (i) yield, (ii) yield and (iii) yield. Here is a very clear demonstration that this is clearly not a good way to conduct a breeding programme. What good is a high-yielding durum wheat if it makes bad pasta? What purpose could a high-yielding sorghum serve if it is effectively lower in digestibility than all other grains? And in a highly polarised debate, both sides of the argument will use the Burkina Faso cotton story to illustrate their points.

Argument A:
Burkina Faso could not sell their GM cotton and the Bt gene led to lower cotton quality and rejection by markets.

Argument B:
The failure of Bt cotton in Burkina Faso was not because of GM, but was the result of poor and hurried implementation by Monsanto.

There is probably an argument C, but that would be from Monsanto. In mid-2017 I searched their website and couldn't really find any mention of Burkina Faso. All I could see was that at the end of 2016 they were releasing Bollgard III, which contains the same Cry genes as Bollgard II plus a new class of insect resistance gene, vip3A. The vip proteins are vegetative insecticidal proteins from *Bacillus thuringiensis* and have a different mode of action to kill insects.

In Australia today, almost 100% of cotton contains two Bt genes, and in most cases varieties have an additional gene for herbicide resistance. Mostly this is Roundup Ready but in some cases there is the so-called LibertyLink technology. This is marketed by Bayer and has resistance to glufosinate ammonia herbicide. Most cotton growers have access to irrigation water, and techniques such as laser-levelling of land, excellent soil nutritional profiling and good well-adapted varieties. Over 1500 families grow cotton – some only grow cotton, but most commonly utilise rotations with sorghum, wheat, barley, mung bean and many also have cattle on the property. Annually cotton is worth at least $1 billion, and in record years, such as 2011/12, can be up to $3 billion. Nearly all the cotton is exported and Cotton Australia proudly proclaims that enough cotton is produced to clothe 500 million people every year. If it wasn't for disposable fashion, Australia could clothe the world. Of course, if it wasn't for disposable fashion the market for cotton would be much smaller.

Cottonseed was transported to Australia in the ships of the First Fleet, arriving in Sydney in 1788. As a crop, it never did too well. Disease, pests, drought, flood and rather poor understanding of management all contributed to the fact that the first export crop left Australia in 1830, a paltry total of three bags, to the mills in Lancashire. Most of the continued experimentation led to small production successes, but the industry never really took off.

The small town of Wee Waa (population ~1600) in New South Wales calls itself the "Cotton Capital of Australia". Apart from the annual cotton harvest in autumn, not much ever happens in Wee Waa, so you probably have never heard of it. French techno group Daft Punk somehow heard of it and launched their new album there in 2013, so maybe if you have *Random Access Memory* you may have heard of Wee Waa. Or maybe you were one of the 4000 people who were there and stayed up all night to *Get Lucky*. Cotton was first successfully grown in the region in the 1960s by a couple of Californian farming families who decided that the Namoi Valley was just what they were looking for when they decided to move from the Merced Valley in central California. Within a few short years, cotton had become the economic mainstay of Wee Waa and the Namoi Valley, especially with the availability of endless water from the Keepit Dam built by the state government in the late 1950s. Ineluctably the locals started to talk about "white gold". Yet another gold rush. The usual Australian weather pattern of "droughts and flooding rains" hit fairly regularly. But it was the arrival of *Helicoverpa* in the mid-1970s that became the true threat to the industry. Cotton farmers were fairly soon spending over $50 million annually to control the pests. So yield was maintained, but profitability was taking a hit.

The other thing taking a hit was the population of Wee Waa, and all the other cotton-growing regions in New South Wales and Queensland. As a university student in the early 1980s, I did a student placement in Biloela in Central Queensland. I worked mostly with the sorghum and cotton breeding teams there. One Saturday I had been invited over to have lunch with the family of the sorghum breeder, an American import, Ray Brengman, now at Kansas State. It was a memorable day, initially interrupted by a brown snake that for some inexplicable reason slithered up to the house. It was not just a brown snake in colour, but an Eastern Brown Snake (Figure 5.2), the second most venomous snake on the terrestrial planet. According to Wikipedia this species is "notorious for its speed and aggression". I grabbed a

Figure 5.2 The Eastern Brown Snake or Common Brown Snake is found on the east coast of Australia, is extremely venomous and can grow up to 7 metres in length.
© Kristian Bell/Shutterstock.

hoe (what an idiot – woefully inadequate and even as a 20-year-old that snake was always two moves ahead of me) and Ray appeared with a shotgun. After a few useless swipes at the snake it went on the attack, whereby I used the hoe as a cross between a pogo-stick and a very short pole vault apparatus, and before the shotgun could be discharged, the snake sought refuge by hiding inside a welding machine that happened to be nearby! At about the same time, the phone rang, and Mrs Brengman shouted, "Quick everyone get the washing in". Seemed like an all-hands-on-deck activity so clothes pegs, clothes, arms and legs were flying in all directions. I had no idea what was happening but assumed that a tornado or a tsunami or some other source of divine retribution was heading our way. Maybe she heard the Eastern Brown Snake's family were on the way?

Finally the washing was in, and only a few items had been dropped in the dirt. Things were now calmer, and as was explained over a beer and a bowl of chilli (a Texan thing that I had not hitherto experienced), and a respectable distance from the welding machine, this frenzy with the washing was a regular occurrence. A cotton-growing neighbour had phoned to say the crop duster was going to be flying over soon to spray his crop for pests. The real point in saying that was to say "you better get the washing in". Good fences make good neighbours but when the fence is not high enough, a phone call is a good thing. Within 5 minutes you could hear the plane and, shortly after, smell the endosulfan, or whatever cocktail of chemicals was being used. This was a ritual occurring daily up and down the cotton region. Depending on the wind direction, the spray drift was something everyone got to smell – up and down the main street, in the pub or milk bar, at school, the cinema or even church. When you can smell something – it is not that you are smelling it from afar. It means actual air-borne molecules from whatever you are smelling are entering your nose and setting off your olfactory senses. So if you can smell molecules, you can be assured that molecules are on your clothes, your skin, in your eyes and hair. It's the same phenomenon you experience when you go to a party/venue/bar filled with people who are smoking, and then next morning you can smell the cigarette smoke on your clothes and hair.

In peak season, spraying was happening every day in any given region, and some growers were spraying their crops up to 17 times per growing season to hold off the voracious caterpillars. That meant weekly in some cases. If you lived in a place like Wee Waa or Biloela, where there were more than 100 cotton growers who were spraying every 2 weeks at least, the problem started to become an air traffic control issue. Crop spraying planes and farmers using tractor-borne spray rigs also had to try to make calculations as to wind speed and direction, and avoid spraying at crucial times – like when the local schoolchildren were playing cricket or netball outside, or walking home from school, or all the locals were at the picnic race day. Or at times when people and wildlife were breathing. It was also a good way to keep the beer gardens deserted. Tooheys and XXXX are ordinary enough beers (opinion, not data-free, but nevertheless an opinion not shared by many Australians) without the added benefit of a *soupçon* of endosulfan.

Bt cotton (Figure 5.3) reduced the need to spray for insects to once or twice per season on average. Hang on – why are farmers still spraying insecticides when these Bt plants are supposed to be resistant? There are two answers to that. The more acceptable answer is that there are other pests that Bt does not affect, and in the absence of spraying broad-spectrum insecticides, "minor" pests have become "major". These include whitefly, aphids, leafhoppers, jassids (Figure 5.4) and mirid bugs, none of which

Figure 5.3 Mature Bt cotton crop at sunset, Emerald, Central Queensland. Photo by Jess Lehmann.

Figure 5.4 Cotton jassid (leafhopper).
© Shutterstock.

are Lepidopteran (moths and butterflies). Whiteflies have been a huge problem in India in some years, and mirid bugs have become a severe cotton pest in some parts of China. Both mirids and whiteflies are serious pests on Bt cotton in Australia in some years. The second answer is that in some years, the insect pest pressure from *Helicoverpa* is so high that insecticide spraying is necessary to reduce the population.

Another feature of Australian cotton production is that the introduction of Bt cotton did not lead to increased cotton yields. This is commonly levelled as a criticism – why would farmers pay for a more expensive variety if it has not improved yield? The criticism is misplaced. For farmers, the end point to profitability is their margin. This is how much did they spend per hectare (seed, diesel, fertiliser, pesticides, labour, water) against how much income they realised per hectare. The income is a factor of yield and price. Increased yield does not always lead to a better margin, especially if world prices go down. However, achieving a better yield is one way to improve the margin, as is producing a higher quality product that results in getting a better price. The other way to increase the margin is to reduce costs, and that has been the big benefit for Australian cotton growers. Reducing the frequency of spraying from 15–17 times per season to 1–2 led to reduced insecticide costs, reduced use of fuel (for the tractor or the crop sprayer) and reduced labour. In other countries where insect management was not as good as it had been in Australia (although relying on chemical insecticides), meant that there were significant average cotton yield gains, like 24% in South Africa, 30% in Argentina, 37% in India and 18% in Burkina Faso.[5]

The other benefits have been environmental and social. Significantly reduced insecticide and carbon fuel usage is one benefit. Not getting spray drift on your clean washing, beer garden or netball game is another. Not having endosulfan and other insecticides in your water supply or backyard salad vegetables is also quite helpful to your physical and mental well-being.

5.1   SOYBEAN AND ROUNDUP READY

Soybean. An ancient crop but a modern phenomenon. Soybean originates from east Asia, with fossil and archaeological records from China, Korea and Japan over 2000 years old. It was used as a minor food crop, with evidence of production of foods like tofu and soy milk, and use in fermented products like soy sauce. Outside east Asia there was little production until World War Two when production in the USA was stepped up as a source of oil and protein for animal feed. The USA was a big importer in 1940, and by 1942 had become the world's biggest soybean producer, where it has remained ever since. Significant advances in agronomy and genetics led to soybean production expanding at an amazing rate, with world production increasing seven-fold from 1949 to 1979, when production exceeded 50 million tonnes. By 1985 world production hit 100 million tonnes, then 200 million tonnes by 2003. By 2017, world production had almost reached 350 million tonnes, with both the USA and Brazil exceeding 100 million tonnes. As well as the high protein and oil content, soybean is a sought-after rotation crop because as a legume, it fixes its own nitrogen from the atmosphere by the wonderful symbiotic relationship with the soil-borne bacterium *Rhizobacterium*.

Soybeans require regular water supply, and as a result are not grown widely in the drier parts of the world like Australia, west Asia or Africa. Other than water, the biggest constraint to soybean productivity has been weed control. For weed control, soy producers needed to use some pretty potent and toxic cocktails of herbicides and/or frequent tillage to reduce the weed seedlings. For taller crops like maize and cotton, weeds would often be covered by the rapid growth of crop biomass. Farmers utilised plant populations and varieties that would produce biomass early and "canopy closure" would put weeds in the shade. Soybean, like other legumes, on the other hand is not a rapid accumulator of early biomass, and weeds are a significant issue, using precious water and nitrogen that the crop otherwise would use. And because the weeds are just as tall as (or taller than) the soybean crop, they compete for sunlight and hence reduce photosynthesis in the crop. Weed control in the first 6 weeks has a major effect on yield. Weed control in the last 6 weeks of the crop can have an impact on ease and cleanliness of harvesting, especially if weeds are taller than the soybean plants.

Before the release of Roundup Ready soybean varieties, farmers used numerous herbicide classes, and there was very little usage of glyphosate (Figure 5.5). Once Roundup Ready (glyphosate-resistant) soybeans were released in 1996, glyphosate usage

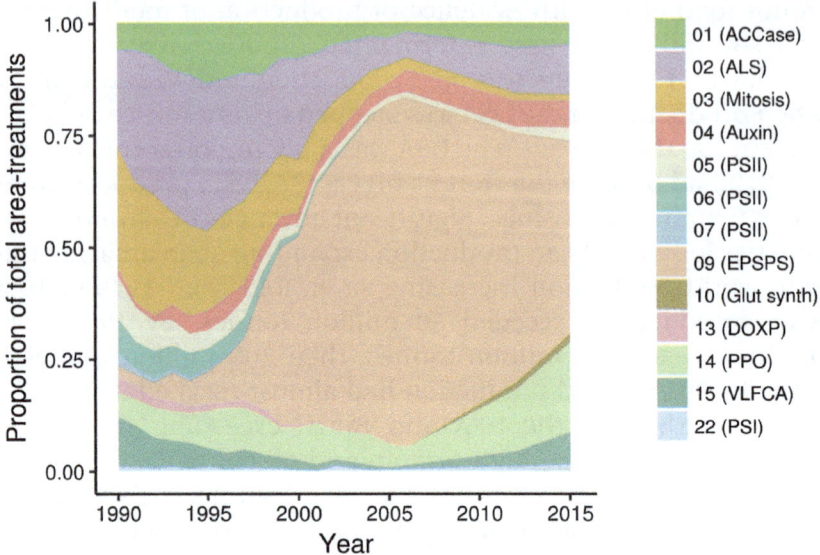

Figure 5.5   Herbicide use in soybean, 1990 to 2014.
Adapted from ref. 6 [https://doi.org/10.1038/ncomms14865] under
the terms of a CC BY 4.0 licence [http://creativecommons.org/
licenses/by/4.0/].

skyrocketed, such that within 10 years it comprised 80% of all
herbicide usage. Andrew Kniss, Professor in Weed Science at the
University of Wyoming, has recently demonstrated that while in
the first 5–10 years this led to a reduction in total herbicide,
this has been reversed in the past decade as farmers started
to use other herbicides in their weed management programs
(Figure 5.5).[1] The major benefit of glyphosate usage was that it was
both a simple management tool and it had a lower toxicity, both
acute and chronic, than most of the other classes of herbicides –
although, as you may be aware, many people have spent many
hours trying to demonstrate that glyphosate does cause cancer and
other chronic illnesses. And there are two reasons for this: (i) they
want to link this to GM plants; (ii) they want to link glyphosate to
Monsanto. But we will talk about these complex issues in later
chapters. This chapter is about what farmers have experienced.

Shortly after Roundup Ready soy was released in the USA, new
varieties became available in Latin America. What happened
next makes a very interesting case study. GM soy was rapidly
adopted in Argentina. In both places, one of the big selling

points was that this enabled no-till agriculture to become a realistic part of the farming system. Growing soybeans without significant tillage (meaning ploughing the soil quite a few times) before planting to remove weeds was extremely attractive to farmers. Why?

1. Tilling the soil costs money (diesel fuel, labour).
2. Tilling the soil a number of times to reduce the seedbank of weeds takes time, which reduces the planting window.
3. Tilling the soil exposes wetter soil to evaporation, hence the soil profile becomes drier, which is not a good scenario when you want to sow a new crop into it.
4. Tilling the soil to remove stubble from the previous crop means there is little vegetation and root mass holding the soil together, so a storm can lead to significant soil run-off into waterways. On more sloped land it can lead to most of the valuable organic matter and nutrient-rich topsoil being washed away. This has a negative impact on the next crop and leads to pollution of nearby waterways (with silt and nutrients).

Time for some more figures. Within 5 years of the release of Roundup Ready soy (by which time over 60% of US soybean was GM), a US Department of Agriculture study[2] estimated the following benefits from the no-till GM soybean system:

- 1.8 million tonnes increase in soybean production.
- A reduction of 12 million kg of herbicide applied.
- Savings of $1.2 billion in insect costs (labour, fuel, pesticide).
- And most surprisingly, a 90% reduction on topsoil run-off.

However, after the first 5 years, herbicide use started to creep up again. And, as predicted by many weed biologists, this was because narrowing the herbicide types that weeds were exposed to intensifies the selection pressure on weeds to become resistant. Or to put it another way, which is more realistic, herbicide resistance genes existed before herbicides were invented. Yet they tend to tick along at fairly low frequencies in a population until they can confer a selective advantage. Hence if on your farm there is one single plant that has a resistance gene to glyphosate, and you start applying glyphosate as your main or sole herbicide, you will kill everything except that one weed. And that one weed, assuming it is self-pollinating, may easily produce 1000 seeds to germinate next year – so next year you have 1000 herbicide-resistant weeds, and the next year a million. So in a standard soybean crop with, say, 100 000 plants per acre, you may not notice 1000 weed plants, but you are definitely going to get a little edgy when you have a million of them, and by then, you have enriched herbicide resistance in your weed population. Some of those other herbicides you said goodbye to 5 years ago may have to become part of your spray regime again. As Matin Qaim told me, he was not a fan of Roundup Ready soybean, not because it didn't work, but because it encouraged poor agronomic practice: "It allowed farmers to become lazy, and that is what led to herbicide resistance issues a few years later." Nevertheless, farmers in the USA continue to use GM soybean, and from 1996 to 2012, the net farm income benefit to US farmers has been $149 per hectare, based on cost saving of low/minimum till and yield gains of 5–11%.[8]

Where the story of GM soybean starts to get really interesting was in South America. Farmers in South America had access to Roundup Ready soy soon after their US counterparts (or competitors as they would see them). The Argentinians embraced the Roundup Ready soybean and the opportunities it gave them. The greatest benefit was not the soybeans themselves, but the opportunity that no-till agriculture offered them meant a shorter growing cycle, hence they could grow wheat and almost immediately plant soybean into the harvested stubble in the same growing season. In Paraguay, many farmers took up the same opportunity. Hence the farm benefit was massive, with a mean $213 extra farm income per hectare just from the second crop.

Over the border in Brazil, it was a very different situation. In 1998, battles led by the eco-warriors at Greenpeace managed to delay official permission to grow GM crops. The Brazilian government decided on a moratorium until 2005, while they waited to see the economic and ecological disaster unfold in Argentina and Paraguay. Or at least that was what Greenpeace had looked into their crystal ball and foretold. They also promoted the idea that the Brazilians could make loads of money by staying away from GM and marketing their soybeans as organic. Brazil was to become the darling of tofu-munchers worldwide. Turns out the Greenpeace crystal ball was a tad myopic, and didn't take into account the Brazilian farmers. The Brazilian farmers, particularly those close to the Argentinian border, were treated to anecdotal and finally hard economic data to show that they were being economically out-competed by the Argentinians. This was a tough pill to swallow. Brazilians and Argentinians have a friendly rivalry, which sometimes becomes distinctly unfriendly, and particularly fierce when they play football against one another.

Brazilians also think that Argentinians are arrogant. A Brazilian student once told me: "We have a saying in South America. The quickest way to get rich is to buy an Argentinian for what he is actually worth and then sell him for what he thinks he is worth."

Looking over the border, it seemed to many Brazilians that the Argentinian soybean growers were becoming richer. So Brazilian farmers took the law into their own hands, and started smuggling in GM soybean seed, ably assisted by Argentinian farmers, seed merchants and truck drivers. By 2003, it was publicly obvious to most people that most of the Brazilian soybean crop was GM Roundup Ready. In the southern state of Rio Grande do Sul, over 90% of the soy grown was GM. Total Brazilian soybean area more than doubled from 10 million hectares in 1996 to 23 million in 2004/05. And farmers absolutely loved it. It was another one of those gold rushes – this time a green one. The streets of most rural towns started to fill with brand-new Ford F-250 trucks, many sporting stickers proclaiming "100% Transgenico". In Julio de Castilhos, deputy mayor Antonia Abreu said "Right now we've got a gold mine. It'll probably become a silver mine, but we can live with silver" (*Daily Herald*, Utah, 21 December 2003). We can probably say that as well as the rural communities, Ford Motor Corporation no doubt loved this new-found wealth too. This was truly an example of a farmer-led revolution and was so unstoppable that the Brazilian government had to concede defeat and agree to end their attempted moratorium in 2003.

Greenpeace was not happy. It was desperate to find herbicide-resistant weeds and show that everyone's health was being adversely affected. The weeds were easy. Health issues? Well let's talk about that later too, but they are still looking. Greenpeace's mortal enemy, Monsanto, were not happy either. Firstly, they were not getting any royalty from the smuggled soybean seeds. To make things worse, even though the use of Roundup had increased markedly in Brazil, the glyphosate patent had expired in 2000. It's not an expensive chemical to make, and hence generic versions had cut the price substantially. Not only was Roundup now not the only herbicide containing glyphosate, Monsanto had to drop their price to remain part of the market. In 1994, about 56 million kg of glyphosate was being used worldwide. By 2014 this had increased to 825 million kg, and during the last few years of the Monsanto patent the market was growing by 20% per annum. From 1994 until the patent ran out in 2000, Monsanto actually dropped the price of glyphosate by more than 45% (www.frost.com), however the price always stayed above $50 per pound of active ingredient. Since 2012, the price has always been below $20.

Soybean stands out as the crop where farmer-led demand for GM varieties has been strong. By 2016, over 90 million hectares of soybean was grown worldwide, which makes up about 80% of the world's soybean production. Most of this is grown in the USA, Brazil and Argentina, where over 95% of all soybean is herbicide-resistant GM varieties. The greatest benefits have been in Argentina and Paraguay, where double cropping has become possible because of no-till production. And in both the USA and Argentina, some farmers have stated that if growing Roundup Ready soybean was fairly cost-neutral (the extra seed costs were roughly equal to the financial benefits), they would continue to use the herbicide-resistant crop. When questioned further, for most of them the greater benefit was in safety for themselves and their families. Changing from a cocktail of herbicides with higher toxicity to the low-toxicity glyphosate herbicide was a personal safety benefit, and many stated that they believed it was safer for their families, who in nearly all cases live on the farm.

Brazilian farmers who led the revolution on growing GM soybean against the moratorium have benefitted financially, but they do have some questions about the ongoing sustainability of the farming system. Although, on average, it has been demonstrated there is a net benefit of $34 per hectare across Brazil,[9,10] which is mostly due to savings in herbicide application (herbicide, labour, fuel). However, the farmers in Rio Grande do Sul experienced some difficulties with their "illegal GM" soy. They were growing lines adapted to Argentina, and were not selected for local adaptation. There was also no local level agronomic extension service advising them, as their operations were "flying under the radar", so to speak. With all those big shiny trucks driving around town with "100% Transgenico" stickers, it's true that everybody knew what was going on, but that doesn't mean that government-funded agronomists and weed scientists could really become engaged in assisting with crop and weed management programmes. That, in itself, could well have exacerbated subsequent problems experienced with herbicide-resistant weeds (Chapter 6). It certainly gives even more credence to Matin Qaim's reservations about herbicide-tolerant crops. Yet it is also worth stressing that there are many cropping systems around with herbicide tolerance traits in the crop, and many of these traits have been developed through "natural" mutations, not transgenics. And the herbicide-resistant weeds don't discriminate between GM crops and mutant-derived crops. They just continue to mutate themselves regardless. As do pathogens and pests.

5.2 GM MAIZE IN THE USA AND AFRICA

The "big green giant" of world agriculture is maize, or corn as the Americans call it. Which does confuse me and many others. In much of northern Europe, corn is wheat, or a generic term for all cereals. As already discussed in Chapter 2, the introduction of hybrid corn and subsequent advances in plant breeding and genomics have hugely increased maize production worldwide, such that in most years, it is the most produced crop worldwide. It is the undisputed number one crop in the USA, and has recently become so in China. Its invasion into Europe, Asia and Africa has been very successful. Total area sown to maize in Germany now exceeds that sown to wheat, although most of this is for silage for biogas production (known as "green maize"). France is Europe's biggest maize producer, and in countries such as Ukraine, Romania, Hungary and Serbia it is the most important crop. In the past three decades, maize has also become the most important grain crop throughout most sub-Saharan African nations. Maize has also become the number one staple crop in many of these countries, supplanting sorghum, millet and cassava.

GM maize has benefitted from both insect and herbicide resistance traits, with the first Bt maize and Roundup Ready maize cultivars released in 1996. Soon afterwards, farmers had access to cultivars with "stacked traits" with both insect and herbicide resistance. The major pests targeted with this technology have been Lepidopteran pests (caterpillars that grow up to be moths or butterflies). The uptake of Bt maize has not been as widespread as in cotton, predominantly because the targeted pests have not been as widespread, and also because there was not such a high level of insecticide resistance among these pests as was found in cotton. Targeted pests are the corn borer (Figure 5.6), which eats leaves and can bore into cobs, and the corn root worm. The corn borer not only causes some cob damage, but as it eats through the protective husk it will often allow the introduction of grain moulds. Some of these cause major health problems with the production of aflatoxins (from *Aspergillus*) and fusiminoforms (from *Fusarium*). Aflatoxins are some of the most carcinogenic substances known. In acute outbreaks they can cause liver disease, stunting in children, and lead to liver cancer. One of the worst documented outbreaks occurred in Kenya in 2003 with 120 confirmed deaths from contaminated maize.

There is conflicting data as to whether Bt corn can prevent aflatoxin contamination in maize. As reported by Windham,[7]

Figure 5.6   The effects of the European corn borer (*Ostrinia nubialis*) in maize.
              © Tomasz Klejdysz/Shutterstock.

while corn borer greatly increase the level of aflatoxin in corn (by 10 to 100-fold), *Aspergillus* infections can occur in the absence of insect damage. It was also reported that even in Bt maize, before the insect larvae stop feeding, they may have already caused sufficient damage to allow the fungus to infect the corn cob. Hence while Bt maize has a role to play in reducing aflatoxin outbreaks in maize, it is not a general panacea.

Overall, on-farm benefits have been impressive. In the USA and Canada, the availability of corn borer (European corn borer and South West corn borer) resistant maize has net benefits of nearly $90 per hectare, whereas the corn root worm resistant corns have net benefits of $89 per hectare and $106 per hectare in the USA and Canada, respectively.[8] Similar benefits were experienced on farms growing Bt corn in the Philippines, Brazil and South Africa. Benefits have been even higher where insect pressures caused more problems, or current control measures were largely ineffective. For example, in Spain and Colombia, the net benefit in farm income was in excess of $200 per hectare. Elsewhere the benefits were considerably more modest, such as only around $20 per hectare in Argentina and Uruguay.

Spain was one of the first European countries to embrace Bt maize. European corn borer had become a problem in Spain, with around 25% of the area under high and 40% of the area under moderate insect pressure. As a result, about 20% of growers spray insecticide on their maize crops to control insects. In 1998, Bt maize became available to farmers in a limited region, and Syngenta agreed to restrict the total area to 20–25 000 hectares, or about 5% of the total maize area. Prior to the Bt maize, most farmers were choosing not to spray with insecticides and tolerating 5–7% yield losses. In areas where insect pressures were highest, farmers were applying 35 000–56 000 kg of pesticides, chlorpyrifos and synthetic pyrethroids. They met with limited success, because often once they noticed the insect, some had already bored through the husk into the cob where the spray could not reach them. This issue of timing was a major reason why many farmers were choosing not to spray but instead to take the yield penalty (around 10%, but sometimes over 20% in Girona and Huesca) and avoid the cost of spraying. The advantage of the Bt maize was that it totally overcame the problem of when to spray. Yield benefits were significant, even after the additional seed cost (sometimes up to $50 per hectare) was taken into account. Within the first year of planting, farmers were enjoying yield lifts of over 10%,[11] similar to those yield gains observed in Illinois and Minnesota.[12]

In addition to the general increase in profitability, there were measured decreases in fungal infection in the Bt maize hybrids grown in both Spain and France. A team from INRA led by Benedikte Bakan[13] showed that *Fusarium* infection and fuminosin levels were orders of magnitude higher on conventional rather than Bt maize. At five sites across France and Spain they detected between 0.3 and 9 ppm fuminosin on conventional maize cobs, yet only 0.05–0.4 ppm on cobs from Bt maize. While they often occur in combination with the more highly toxic aflatoxins, it can be difficult to assign health problems only to them. However, the family of fuminosins has been demonstrated to be associated with neural tube defects and oesophageal cancer.

Yet none of this stopped groups like Greenpeace from stepping in and demanding that the "toxic" GM crops should no longer be grown. It seemed that they had a strong preference for the conventional toxicities of aflatoxin and fuminosins on their maize, although that was never stated in their political campaigns. However, by 2006, the area sown to maize increased to over 53 000 hectares, and made up about 60% of the maize area in the worst areas for insect infestations. By 2012, the area of GM maize exceeded 100 000 hectares and has stabilised to be around 30% of all maize in Spain. In France, it was a different story. In 2008 the French government banned Bt maize.

Like Bt cotton, farmers have been required to grow refugia of non-Bt maize. This maize must make up a minimum of 20% of the total maize area on each farm. Growers have been very diligent at ensuring these refugia are grown, and using integrated pest management have been able to grow Bt maize without the development of resistance. This has surprised some scientists as the target insect, *Sesamia nonagrioides*, has now been exposed to the same Cry1Ab toxin for almost two decades. It may be that the slow introduction and gradual increase of Bt maize has slowed the development of resistance, however some modelling has predicted that Bt maize may continue to grow in Spain for another 20 years without any appreciable levels of resistance.[14]

5.3 GM "RAINBOW" VIRUS-RESISTANT PAPAYA

We've already visited Hawaii and the South Pacific to discuss the miracle of the Rainbow papaya, overcoming the scourge of papaya ringspot virus. It's not so much that the virus-resistant papayas helped improve yield and quality, they actually saved an industry from total collapse, as had been already experienced over 50 years ago on the island of Oahu. Similar local industry collapse has occurred in parts of South East Asia and many of the islands of the South and West Pacific, and papaya ringspot virus remains the largest constraint to papaya production in most parts of the wet tropics, where papaya is an important part of local diets. Farmers are advised to use control measures, mostly aimed at the aphid vectors. Mostly this means insecticides on a monthly basis, and applying silver reflective plastic to the soil. These measures reduce the in-crop infection, but generally whole regions have to agree not to grow papaya for a number of years to eradicate the infection. This has been used in the Philippines, Thailand and Brazil to attempt disease eradication. In both the Philippines and Thailand, trials with locally produced GM virus-resistant papayas have taken place.

Yet, at least officially, no jurisdiction other than Hawaii has allowed for the introduction of the Rainbow papayas. What is going on? What are the forces at play behind GM moratoria, and the refusal of some jurisdictions to allow any GM crops?

REFERENCES

1. T. Vilsack, *Budget Hearing – Office of the Secretary, Department of Agriculture*, Thursday February 11, 2016, Agriculture, Rural Development, Food and Drug Administration, and Related Agencies, The U.S. House of Representatives Committee on Appropriations, https://appropriations.house.gov/calendararchive/eventsingle.aspx?EventID=394359.
2. A. Subramanian and M. Qaim, The Impact of Bt Cotton on Poor Households in Rural India, *J. Dev. Stud.*, 2010, **46**(2), 295–311.
3. J. Kathage and M. Qaim, Economic impacts and impact dynamics of Bt cotton, *Proc. Natl. Acad. Sci.*, 2012, **109**(29), 11652–11656.

4. J. Vitale and J. Greenplate, The role of biotechnology in sustainable agriculture of the twenty-first century: the commercial introduction of Bollgard II in Burkina Faso, in *Convergence of Food Security, Energy Security and Susutainable Agriculture, Biotechnology in Agriculture and Forestry*, ed. D. D. Songstad, J. L. Hatfield and D. T. Tomes, Springer-Verlag, Berlin, 2014, vol. 67, pp. 239–293.

5. G. Brookes and P. Barfoot, Global income and production impacts of using GM crop technology 1996–2014, *GM Crops Food*, 2016, 7(1), 38–77.

6. A. R. Kniss, *Nat. Commun.*, 2017, **8**, 14865.

7. W. Paul Williams, G. L. Windham, M. D. Krakowsky, B. T. Scully and X. Ni, Aflatoxin Accumulation in BT and Non-BT Maize Testcrosses, *J. Crop Improv.*, 2010, **24**(4), 392–399.

8. P. Barfoot and G. Brookes, Key global environmental impacts of genetically modified (GM) crop use 1996–2012, *GM Crops Food*, 2014, 5(2), 149–160.

9. A. Galveo, Farm survey findings of impact of insect resistant corn and herbicide tolerant soybeans in Brazil, Celeres, Brazil, 2010, Available from: www.celeres.co.br.

10. A. Galveo, Farm survey findings of impact of GM crops in Brazil, Celeres, Brazil, 2012, Available from: www.celeres.co.br.

11. E. Alcade, (1999) Estimated losses from the European Corn Borer, Symposium de Sanidad Vegetal. Sevilla, Spain.

12. M. Marra, P. Pardey and J. Alston, *The Pay-offs of Agricultural Biotechnology: An Assessment of the Evidence*, International Food Policy Research Institute, Washington, USA, 2002.

13. B. D. Bakan, D. R. Meleion and B. Cahagnier, Fungal growth and Fusarium mycotoxin contention, *Isogenic Traditional Maize and Genetically Modified Maize Grown in France and Spain*, 2002, vol. 50, issue 4, pp. 728–731.

14. P. Castañera, G. P. Farinós, F. Ortego and D. A. Andow, Sixteen Years of Bt Maize in the EU Hotspot: Why Has Resistance Not Evolved?, *PLoS One*, 2016, **11**(5), e0154200.

CHAPTER 6

Bad Moon Rising

"Nothing is so painful to the human mind as a great and sudden change"
                    Mary Shelley (Frankenstein, 1818)

"Those who can make you believe absurdities, can make you commit atrocities"
                    Voltaire

In the early 1990s, with my plant pathologist colleague, Liz Aitken, we attempted to convince the Australian banana industry that it was an essential long-term strategy for them to invest in banana genetic improvement. The local industry (like most others around the world) was based almost entirely on a single cultivar, the Cavendish banana. If you know your bananas, you will also know that the Cavendish, while a worthy workhorse, is not exactly the most flavoursome banana around (Figure 6.1). In addition to that, Cavendish bananas are vulnerable to some pretty savage fungal pathogens, including a couple of foliar pathogens known as yellow Sigatoka, and black Sigatoka, after the Sigatoka Valley in Fiji where they were first recorded. These pathogens are pretty lame though, because while they are economically troublesome, they can be controlled by regular application of fungicides and sanitation measures. This however

Good Enough to Eat? Next Generation GM Crops
By Ian D. Godwin
© Ian D. Godwin 2019
Published by the Royal Society of Chemistry, www.rsc.org

Figure 6.1   Various varieties of banana: red, Pisang Raja, Cavendish (photo by
            Andre Drenth).

is not the case with the biggest threat to banana production
worldwide.

The deadly Panama disease, caused by the fungus *Fusarium
oxysporum* f.sp. *cubense* (*Foc*) has already had a distinguished
history of being an industry-changing disease. There are so many
diseases caused by *Fusarium oxysporum* that taxonomists have
opted to differentiate them with a taxonomic device known as
forma speciales (that's the f.sp. part of the name). So, there are
forma speciales lycopersici (a big disease of tomato), *asparagi*,
cannabis, *citri*, *lini* (flax), *batatas* (sweet potatoes), and almost
30 others. How exactly that taxonomic device came up with
cubense while the common name is Panama disease is one of
taxonomy's little mysteries. Cuba is only 1472 km from Panama
as the crow flies, but it is a separation entirely promulgated by
open sea. Still, maize rust remains *Puccinia sorghi*, even though it
has never infected sorghum, and *sapiens*, as in *Homo sapiens,* is
supposedly Latin for wise!

So being sapient, and standing on the shoulders of others, I
took to the wonderful internet to be informed that the disease
came neither from Cuba nor Panama, but is thought to have
co-evolved with the host in the Indo-Malay Peninsula. "Host" is

pathologist talk for the banana – they cease to be plants, and instead become hosts when the fungus is your favourite part of the interaction. However, as a disease the first recorded observation was in the 1890s in Panama and Costa Rica in plantation bananas. By the early part of the 20th century, most of the world's traded bananas were of a single cultivar. That single cultivar was called Gros Michel (literally Fat Michael). Unfortunately, Gros Michel turned out to be susceptible to Panama disease, with devastating consequences. Between 1950 and 1960, Gros Michel was largely replaced by the Cavendish banana. The Cavendish was resistant to Panama disease, and by virtue of its thinner skin making it more difficult to transport without bruising, as well as being considered lower eating quality, it soon become the predominant banana around the world. Today, the Cavendish makes up over 90% of internationally traded bananas. So for many of us, particularly in Europe and North America, the Cavendish banana is "how bananas taste", because quite often that is the only banana available in the supermarket. That notwithstanding, there are many tastier bananas around the world, including a popular favourite (and mine) throughout South East Asia, Pisang Mas (the golden banana). Pisang Mas is a thin-skinned small banana, which falls off the bunch when ripe so it does not lend itself to travelling great distances to market. However, a guide to bananas published in *The Straits Times* newspaper (15 April 2014) in Singapore put forward the considered opinion that Pisang Raja (the king banana) is the best and most flavoursome banana. The same article reported that Cavendish bananas have "creamy but bland flesh" and are "fragrant only when extremely ripe", which is when many Australians throw them out. They start to smell like a banana, and many Australians identify that aroma as over-ripe or maybe even rotten.

Now, like many pathogens put under the selection pressure of monoculture, the *Fusarium* pathogen has mutated and is coming back more strongly than ever. The Panama disease that destroyed Gros Michel was known as race 1. Since then, we have now got up to race 4. To add to the taxonomic clarity, race 3 doesn't even infect bananas, it only infects the related ornamental *Heliconia*. To add to the confusion, there are two race 4 types, a sub-tropical and a tropical. Tropical Race 4 (TR4; Figure 6.2) is the biggest and baddest of them all, the Sauron of Foc.

Figure 6.2   A banana tree with symptoms of fusarium, FoC TR4, in Tanzania
(photo by Elizabeth Aitken).

Cavendish is susceptible to TR4, as are most cultivated bananas
and plantains including Gros Michel, Pisang Mas, Bluggoe,
Silk and Pome. TR4 has already wiped out huge banana-growing
plantations in Indonesia, Malaysia, China, Taiwan and the
Philippines, and has been recently reported in Laos, Vietnam,
Pakistan, Oman and Mozambique. Note that Mozambique is
way, way across the Indian Ocean (over 8000 km from Indonesia)
in southern Africa. The fungus is soil-borne and enters through
the base of the plant, grows up the vascular tissue of the plant,
eventually causing wilt symptoms because the fungus has
blocked much of the xylem. Xylem are the vessels that carry

water and nutrients from the soil to the top of the plant. A single xylem vessel is continuous from the root tip to the top of the plant, meaning in some trees they are over 100 metres long. But as you know, banana is a herb not a tree. By the time the wilt symptoms are seen in a banana plant, it is too late and the infected and symptomatic plant is already at death's door. But wait, there's more. Banana is cultivated asexually, usually by the means of cutting young developing "suckers" from the base of the plant. These suckers may look healthy (be asymptomatic), yet they may already be infected with the *Fusarium* fungus. Hence infected asymptomatic suckers are the most common way for the disease to spread. This can be especially devastating when a commercial nursery is (unknowingly) sending out shipments of planting material that carry the infection. And it may also be carried in soil, attached to boots, digging implements or farm vehicles.

But wait, there is even more. The major means by which fungi reproduce and spread is through spores. These fungal spores are everywhere in the atmosphere, and there has been ample demonstration to show that some of the lighter spores can, given the right conditions, spread vast distances in the wind. So it's not just that a storm can carry infective fungi across continents like Europe and North America, they even have intercontinental capabilities. New strains of fungus can actually spread across oceans, such as across the Indian Ocean from Africa to Australia. Or potentially in the other direction, as was very probably the case of coffee rust spreading from Africa to Brazil.

Perhaps the worst part of the considerable armoury of *Fusarium oxysporum* f.sp. *cubense* (Foc) is that it can form two quite different types of spores: the common or garden variety conidiospores, and the less frequent but decidedly sinister chlamydospores. Chlamydospores are characterised by having very thick cell walls, and are capable of surviving long periods of hot and dry conditions, which is a pretty nifty adaptation when you live in the soil in the tropics.

So back to the 1990s. In a meeting with the banana industry back then, they were not too positive about investing in breeding and genetics. Most of their research levy was being put into quarantine. This was important to keep out the diseases (which pretty much surrounded Australia to the north, west and east), and perhaps more importantly, to keep imports out. Australia has a proud record of using quarantine to keep pests and disease at bay, but industries do tend to overuse them to put up trade barriers. Among the banana growers, many felt there were ships full of bananas from the Philippines or Ecuador sitting just outside the Great Barrier Reef waiting for the diseases to spread to Australia. Hence the banana industry was investing about $2 million a year on quarantine, and regardless, yellow Sigatoka had already arrived in North Queensland. "What will you do if and when black Sigatoka and TR4 get into Australia and you can't grow Cavendish anymore", I asked the chair of the Queensland Banana Growers research committee, using my most incisive inquisitorial skills. Without even blinking he replied, "I'll just go back to being a cane grower, sonny".

Within a few years, black Sigatoka had arrived ("a number of incursions were reported") in North Queensland. This disease can cause yield losses of more than 50%, mostly because it leads to rapid loss of leaves. As with all plants, this leaf loss is generally a downward slide towards much lower biomass due to the loss of photosynthetic leaf area, meaning the plant can't intercept as much light energy as it could with its full complement of leaves. In places such as Central America where the disease has become entrenched, the costs of aerial spraying are estimated at 15–20% of the final market value of the fruits.[1] Current practice usually includes spraying the entire plantation with a class of fungicides known as sterol demethylation inhibitors 25–40 times per annum. These fungicides interfere with the ability of the fungus to make cell membranes, which is generally terminal for the fungus.

Then in 1997, Panama TR4 was detected in northern Australia, just outside Darwin. Unlike the Sigatoka diseases, there is no known control that allows continued banana cultivation. Fungicides do not work. Soil fumigation has only a transient benefit, because the pathogen always returns. Another effect of soil fumigation is that all those wonderful "good" bacteria and fungi in the soil are wiped out too. At present, the only known solution is to quarantine the area, prevent movement of bananas and plant material from the quarantine zone, and never grow bananas in infected soil again. That was the solution in the Northern Territory, and not just commercial crops, but backyard banana plants were all destroyed by government order. I asked my University of Queensland colleague, Professor Liz Aitken, about the disease. Liz is part of an international network of scientists tackling the slowly spreading disaster that is TR4.

Liz told me that TR4 was a very aggressive strain of the disease. In fact she used two terms to describe it: "bizarre" and "sneaky". Unlike the foliar disease, black Sigatoka, it is impossible to eradicate *Fusarium*. She then let me into a little secret "hot off the press" which I now have permission to share with you. All bananas have *Foc* growing inside them as an endophyte. An endophyte is a microorganism (bacterium or fungus) happily living inside a plant, very commonly in the spaces between cells. In fact it is the most common endophyte of banana, and appears in most cases to be totally benign, and may even be beneficial. However, sometimes it goes bad. If you want to think of a similar human-centric example think *E. coli*. Sometimes *E. coli* goes bad, and has a long history of causing serious diseases and death, such as the outbreak in northern Germany in 2011 which resulted in 53 deaths and almost 4000 instances of food poisoning from organic fenugreek sprouts.

After being in the Northern Territory for almost 20 years, in 2015 the disease was detected on a farm in tropical North Queensland, right in the heart of Australia's most productive banana farms. To drive from Darwin to Tully is over 2700 km, but only 1700 km as the spore flies. Over 90% of Australia's banana production is in this region. The response of the banana industry was swift(ish). The farm was closed down and the land was cleared of bananas, and eventually the Australian Banana Growers Council, with some government assistance, bought the 150-hectare property. The official line is that no bananas will be grown on the property for the next 40 years. Liz told me the best thing they could do would be to let rainforest regenerate on the farm. In addition, it needs to be made quite secure, as animals like feral pigs could spread infected soil from the area. In live bananas, the fungus will move up the plant and sporulate on the leaves, and these spores can spread far and wide to make new infections. But just getting rid of the live bananas is only the first step.

"It's such a sneaky fungus and it's good at everything. If there is no live banana plant it can survive as a saprophyte, produce

chlamydospores and seems to be able to survive for years quite happily on other plants, probably on the roots. You can eradicate black Sigatoka, just get rid of the infected leaves. With Fusarium, you will never eradicate it. Some of the areas that succumbed to race 1 in the 1960s, if you plant susceptible Lady Finger bananas on that land now, they will succumb to the disease within a year. In Taiwan they have turned the crop into an annual. They plant, get fruit for a year, and then get rid of the plants and replant. It's very expensive and probably not sustainable."

Professor Liz Aitken, University of Queensland

Within 2 years, the incursion had spread to the MacKay family enterprise, the biggest banana-growing group in Australia. Bananas are Australia's largest horticultural industry. Over 5 million bananas are eaten every day, meaning one out of four Australians have a banana every day. The MacKay family has been growing bananas in North Queensland for over 70 years – Stanley MacKay planted the first bananas in the Tully Valley in 1945 and today they employ a workforce of 550 people, two-thirds of whom are backpackers. They are also industry leaders when it comes to efficiency and innovation. One of their latest innovations is placing fresh banana vending machines in the streets of Brisbane. Stanley's great-granddaughter, Nicola MacKay, is currently studying a science/agribusiness degree at UQ, and has been doing research on TR4 in the Aitken lab. She explained to me that TR4 first came into Australia the year she was born, and only made it over to Queensland when she started studying plant science at university. She had heard of TR4, and had some awareness that it was a threat to their business, but it really wasn't until it made it to the Tully Valley that the threat became real. The MacKay family has two farms directly adjacent to the infected property, one upstream and one downstream. They had always assumed the scenario that the disease would most likely spread downstream, but they instigated biosecurity measures on both farms as good practice. No vehicles could leave or enter the properties, and the same for all farm implements and boots. Once a boot, tractor or knife was on one of the farms, that's where it stayed.

Nicola was actually home in the mid-year university break in July 2017 when the first infected plant was discovered on their largest farm. The farm workers had been trained in what to look for, and when one of them spotted a suspect tree he reported it to Nicola's uncle, Gavin. Gavin was pretty sure that the tree had TR4 Panama disease and straight away called Biosecurity Queensland and the MacKays instigated their quarantine procedures then and there. The area around the infection was cleared, treated with urea (which is supposed to reduce the fungus's viability), and covered in plastic. Nicola took part in the whole biosecurity process wearing a white disposable biohazard suit.

The MacKay family owns and operates properties from Cooktown to Bundaberg. To put that in a context you may find more illustrative, driving from Cooktown to Bundaberg in Queensland is about the same as driving from New York to Florida, just under 2000 km. Or you could drive from Hamburg to Paris and back, unless you are driving from Hamburg, Pennsylvania to Paris, Texas, then it's only one way.

As a university teacher, I get to ask students questions every day (and *vice versa*). It's not often, however, that I get to ask a student "what's the best and worst-case scenarios" of the effects of TR4, and it's not a hypothetical. It's real, it's very visceral, and it is a very personal question about the future of a family agricultural activity that has become a big enterprise. Nicola was not phased at all. "I think biosecurity is everything – we can't stop it but we can slow it down. We've also gone through geographical diversification and product diversification – as well as bananas we produce sugarcane, cattle, papayas and cocoa". However, they would not like to see the banana industry disappear, and they are supporting some traditional breeding to attempt to find some long-term resistance to the disease. And given that Nicola is a plant science student, she has already been made aware of CRISPR and the new gene editing technology (see Chapter 9 for details).

"We know conventional breeding may not be feasible, and banana breeding is slow and hard because of sterility issues. CRISPR would be good if that can be used to develop a form of resistance, and I guess GM bananas may be a last resort. We would have to take into consideration the potential for adverse reactions of consumers to GMOs."

There is no doubt that GM is an extremely important marketing issue. One of the MacKay's most recent new products is a red-fleshed papaya, Reblo. This is a cultivar they developed themselves and on their website (reblo.com.au), they include an important piece of information "Reblo is NOT a GMO". Without a doubt, this is important to many consumers. The devastating papaya ringspot virus (PRSV) is present in Queensland, but only in the south, hence the MacKay's current papaya growing activities are more than 1500 km north and currently PRSV-free. Hopefully, good biosecurity will prevail and the virus will not spread north. There is no doubt, that like all other papayas, Reblo will succumb to PRSV if it makes a move to North Queensland.

The Hawaiian papaya industry, as we discussed in previous chapters, was saved by the availability of GM PRSV-resistant varieties. The Hawaiian papaya industry was not huge in economic terms, with farm-gate production of around $16 million from around 120 farms. When PRSV came to Hawaii in the 1990s, these farmers could have just stopped growing papayas and the USA could import from elsewhere. The farmers could have changed their farming practices and grown something else. Or like the Hawaiian sugar producers, they could have just sold their land off to developers. Where sugarcane was once grown, most areas have become hotels, golf clubs and resorts, and the very last sugarcane crop was harvested in Hawaii in 2016.

Yet the papaya industry persisted. They embraced the new GM technology to develop suitable new virus-resistant lines, and it paid off. In 2017, over 80% of the papayas grown in Hawaii are GMOs, mostly of the highly successful Rainbow variety. This shining example of how the new GM technology had saved an industry is not particularly well known. Maybe that's because it was successful. Not surprisingly, it was so successful that other regions of the world with PRSV problems looked to Hawaii for inspiration. While never a dominant crop in any part of the world, papaya are important in both local markets and as a subsistence crop among some of the world's poorest people. They are low-maintenance small trees that are easy to grow, and about 75 g of the fruit provides the recommended daily intake of vitamin C.

In other regions where PRSV was adversely affecting the papaya industry, such as in the Philippines and Thailand, national programmes were initiated to get PRSV-resistant versions of their own locally adapted varieties. Not only was local adaptation important, it was of paramount importance to ensure that the local varieties for which the resident populations had a taste preference were altered to give them PRSV resistance. Within a few years, locally adapted PRSV papayas had been developed for farmers and markets in Thailand, China, the Philippines, Brazil, Jamaica and Venezuela. Local trials started to test the plants in the field to assure their yield, virus resistance and quality. And that's when the problems started.

Sarah Evanega is a Professor at Cornell University in upstate New York, and currently Director of the Cornell Alliance for Science. The Alliance for Science was established in 2014 with funding from the Bill and Melinda Gates Foundation, with the mission to promote access to scientific innovation to enhance food security and improve environmental sustainability. They promote specific projects, run short training courses and somewhat longer Global Leadership Fellowship Programs. Among their initiatives is "Embracing GMO Science". Sarah undertook a PhD in which she studied the acceptance (or otherwise) of the Hawaiian GM papayas. In 2008 she wrote an excellent paper "Forbidden Fruit" which was published in *Plant Physiology*, the journal published by the American Society of Plant Scientists since 1926.

Her study starts with the activism against the GM papayas in Thailand, which she witnessed at first hand in 2004. She believes that this was a real turning point, not just for Thailand but all of Asia. Thailand had been a regional leader in embracing biotechnology for food security, human livelihoods and human health. What she witnessed stunned her, and her opening paragraph really captures what was happening in Thailand, and was being replicated all over the world.

*"Dressed in white, hooded "personal protection suits," Greenpeace activists donned goggles, gloves, and respiratory masks—the kind of dress you expect to see in the clean zone of a nanotechnology laboratory, not in a field in bucolic northeast Thailand. Easily bridging a barbed wire fence with a stepladder, they began pulling transgenic papaya (*Carica papaya*) from the trees, throwing the fruit into biohazard waste bins.*

The protestors stood for photographs—the press had been alerted—before a large yellow banner printed both in Thai and English that read: "Stop GMO Field Trials."

Sarah Evanega, Plant Physiology[2]

The astute reader (and you must be, you've made it to Chapter 6) will have noticed a few things here:

- Greenpeace
- Goggles, gloves and masks
- Biohazard waste bins
- "the press had been alerted".

These were not the actions of an enraged groundswell of public opinion with spontaneous actions. These were not the result of a reasoned debate on National Public Radio or the BBC. These were not farmers protecting their livelihoods. No, these were actions designed to make a great visual impact and excellent TV. This was a campaign. Give a TV news channel the opportunity to interview a scientist talking about the wonders of virus resistance. Then give them an activist in a HazMat suit killing plants before those toxic "triffids" start walking to the local pre-school to poison small children. Well it's a pretty easy choice isn't it? Why stand around and have a reasonable argument with somebody when you can just punch them in the nose?

Who was being punched in the nose? Well there's a long list. I can tell you for one that the plant science community felt like we were taking a hit. At first, it was laughed off. It was like when you go to a barbecue and some 5-year-old punches you as a welcome. You laugh it off, despite the painful reality that he punched you in the groin and the laughter is mostly to hide the tears, and suppress the knee-jerk reaction to respond with violence or profanity. To sum up the *zeitgeist* in the scientific community, we were rather incredulous, our noses hurt, and the tears were real. Just like the 5-year-old who, while possibly not well brought up, didn't mean it and didn't really understand what they were doing. "We're doing good stuff here", we thought. These people aren't really against our wonderful science. We are here to save the world – who could possibly question our motives? Who could possibly fail to see that all this wonderful genetics stuff is the way of the future and the next green revolution? Turns out the activists were punching us on the nose with much the same motivation – they were here to save the world too. Made it rather awkward that their version of saving the world involved destroying the botanical outcomes of our version of saving the world. It did rather put a dent in our *esprit de corps* to say the least. However, as we will see in Chapter 7, the punch was not really aimed at our noses, it was firmly targeting the political establishment and those large multinational corporations with their chemical hearts. Just a shame we plant scientists were like Ronald Reagan, we "forgot to duck".

> *"Papaya is being devastated and we have a solution right here. It all comes down to political will. If you want to have impact, you have to be political. That is the essence of modern life."*
> Dennis Gonsalves

In his excellent 2017 film *Food Evolution*, Scott Hamilton Kennedy started the film with the Rainbow papaya story, with archival footage and up to date interviews of the main protagonists in Hawaii. On one side (the ayes) were the papaya farmers and the scientists, on the other side (the nayes) were a number of liberal members of the county council on the "big island" Hawaii, who are drawn to the promise of an organic future. The proposal to ban GMOs from Hawaii was championed by council member Margaret Wille, who drew in a number of "experts" to educate the council regarding the dangers posed by GMOs. Utilising a groundswell of public opinion and urged on by a number of people with not particularly scientific opinions, the discussion soon became a litany of the downfalls of using GMOs. As *The New York Times* put it (Amy Harmon, 4 January 2014), some of the experts had no credentials, and those who did have the credentials were cast as pawns or "shills" of big biotech companies.

Chief among those convincing the local council of the evils of all GMOs was Jeffrey Smith. Jeffrey Smith has written anti-GM books and directed and produced an anti-GMO movie, *Genetic Roulette*. However, he didn't necessarily come across so convincingly in *Food Evolution*, where he didn't have control over the script and direction. In *Food Evolution* he can be seen addressing the Hawaii council public hearings *via* video link, and stating that eating GM papaya will make you more susceptible to colds, flu, hepatitis and HIV. He also appeared on a number of TV shows in the USA making similar claims, without any evidence (scientific or otherwise).

Professor Bruce Chassy, a biologist who was then the head of the Department of Food Science at the University of Illinois, publicly dismissed these claims, pointing out that Smith's only professional experience was as a ballroom dance teacher, yogic flying instructor and a political candidate for the Natural Law Party.[†] Poignantly, or perhaps hilariously (I find it hilarious), Smith refuted this brief resumé in the *Food Evolution* documentary by making the correction that he was actually a swing-dance teacher. I have little to add to this, except to say that I have no personal opinion as to which type of dance qualifies a person to be a more plausible expert witness on GMOs. I will be the first to admit that I dance like Peter Garrett from *Midnight Oil* (if you have seen it, you remember it, it's dancing but not as we know it). Garrett was Australian Environment Minister (2007–2010), so perhaps he would have a better idea?

Agriculture changes all the time. It's really only in a subsidized place like Europe that farmers can grow the same crops year after year, or have the same rotation that their father and grandfather had before them. Well, no, actually that's not even true in Europe anymore. Modern agriculture is dynamic, and markets, politics and human behaviour dictate what farmers can grow, and what they can make money from producing.

In the 1960s, maize was a relatively minor crop in places like Germany, and in 1964 less than 20 000 hectares were harvested. Fifty years later in 2014, this had increased to almost 500 000 hectares. However, the real reason why you see maize everywhere as you speed along the autobahn is because of the "green maize" area of 2 100 000 hectares. This maize is not grown as a grain crop. The whole above-ground plant of green maize is harvested to make silage for cattle feed, or put into fermenters to produce biogas to generate electricity.

[†]The Natural Law Party was founded on "the principles of Transcendental Meditation" based on the teachings of the Maharishi (the self-styled "great seer").

There used to be a dairy farm in my Brisbane suburb, only 6 km from the city when I was a kid, but now there are barely any dairy farms in south-east Queensland. In 1960 there were almost 1.2 million dairy cattle in Queensland. By 2014 there were only 175 000. My own children's primary school used to be a pineapple farm when I was a kid. Nobody in Australia grew canola (in fact it didn't exist until the 1970s) and only immigrants from Mediterranean backgrounds regarded olive oil as something fit for human consumption. Now, it would be hard to find an Australian household without it.

The introduction of GM crops into Hawaii came at a time when great change was occurring in agricultural production patterns. The sugarcane and pineapple industries had collapsed. For most Hawaiians, these had been fixtures of the landscape. These industries collapsed because they could not be produced in an economically viable way any longer. These farmlands had once sustained livelihoods and through countless movies and the travel of tourists over many generations, people identified these landscapes with these tropical crops and associated them as being quintessentially Hawaiian. And what could be more Hawaiian than the macadamia nut (which is actually native to SE Queensland, Australia), the pineapple (from Brazil), taro (from South East Asia) and sugarcane (from Papua New Guinea)?

Changes in the rural landscape can sometimes occur without the urban or peri-urban population taking much notice. It's most noticeable when a sugarcane field disappears and a golf course is built in its place, for example. However, it can also be quite noticeable when a perennial crop (sugarcane, pineapple or macadamia) is replaced by annual crops, like maize or soybean. Seed companies had been using Hawaii as a base for seed production for many years. It had some significant advantages which included:

- A modern, well-educated workforce.
- Good access to transport (land, air and sea).
- No winter. In fact the temperature year-round in Hawaii was around a mean of 24 °C, and even in the winter season, the days don't get shorter than 11 hours, and the average temperature varies from 23 to 27 °C. Hence it is possible to grow annual "summer" crops all year, and up to three crops per year on the same field are quite feasible.

- The region is away from the main crop production areas (for the USA, the Midwest and the South). This meant that there was less likely to be problems of pollen contamination of production crops, and also different pest and disease pressures.

So if you once lived next to a sugarcane field, and it suddenly changed to a maize field, you were going to notice the change. The sugarcane would sit there with fairly minimal management intervention for much of the year, with an annual harvest. The predominant use of these fields had changed to the production of seed – seed for planting in the main agricultural areas. And most of those companies discussed in Chapter 4 were involved. So, now there were many more and frequent agricultural activities. Crops were sown and harvested three times per annum. They were sometimes sprayed with herbicides and fungicides during the cropping cycle. And when crops were harvested, fields were left bare (fallow) for some weeks. On windy days, dust blew over local houses and schools. Some of the seed production was GM. Given that most of the maize and soybean for the US markets was GM seed this was not surprising. So it wasn't the cropping system change but the introduction of GMO that became a focal point. Where, once, a local farmer was producing sugarcane for the now defunct sugar industry, here we had multinationals with chemical hearts producing GMOs and spraying chemicals all over them. Here were some evident changes in the agricultural landscape of Hawaii, and it was not a difficult matter to get local people upset about these changes. In the 1950s and 1960s, the sugar industry underpinned the Hawaiian economy, and it was the biggest employer in the state. Productivity was going up, but the number of farms was going down. In 1990, the farm-gate value of sugarcane was $213 million, yet by 2001, this had declined to $58 million. By 2010, the seed industry was worth around $250 million annually, far exceeding the value of all other agricultural produce from Hawaii. But in the minds of many it was run by big multinationals, and the public equated it with GMOs.

As Zen Honeycutt put it, "GMOs equals pesticides". Zen Honeycutt is an activist who leads a group called Moms Across America. She is very much focussed on campaigning against the

evils of big biotech companies and "contamination" of food with
GMOs. Many of her public utterances start with "As a
mother...", and without a doubt she subscribes to the adage
"mother knows best". It's another variant on the theme of
"Believe me ..." or "The reality is ...", which politicians use to
great effect – or so they like to think. You can believe their reality,
just don't try and do your own research. Or if you really must do
your own research, just use their website. Oh, and while you're
there, there is some really good product available through their
website that you can really trust. It has a sticker on it saying it is
GMO-free. Even though there are no GM oats in commercial
production. Marketing tests will show that the sticker will make
you part with a little more cash because you know it's better for
you than the packet beside it. The packet beside it that doesn't
have the sticker, but may well contain oats from the same farm
and factory.

The *Food Evolution* documentary also uncovers some hilariously candid statements from anti-GM activists. Zen Honeycutt
spoke about trust:

> *"Frankly I trust the social media ... like other Moms that do a
> post. I trust what they say, more than most medical doctors,
> more than the CDC, more than the FDA, more than the USDA,
> more than the EPA. That's real. I don't need a scientific study.
> I don't need a doctor."*

So here we have a problem, do we not? In the world of Fake
News, we have experts who are only self-proclaimed. Then there
are experts (the "shills") who have only said what they are saying
because they have a corporation telling them what to say, and
governments that can make things up as they go along. Taking
their leadership from the leader of the free world (at the time of
writing, President Trump), every fact demonstrating the opposite
can be dismissed as Fake News. You can deny things that were
said, even when there is video evidence to the contrary: "That
was just locker room talk".

Where did the angst about GM crops and foods come from?
Were there reasons to be worried? The answer to that is, without
a doubt, yes. There have been some well-documented cases
which anti-GMO activists could cite to show that this was bad

and dangerous technology. So now we will explore some case studies of where GM techniques have been used to develop some products that did not live up to their promise, or did not go as planned, or were just plain stupid from the original conception of the nucleus of the singularity of an idea for the project. Some plans are cunning, and some are best forgotten. So let's find out about "What could possibly go wrong?"

6.1 HIGHLY NUTRITIOUS SOYBEAN

If you are a human, or a pig (you be your own judge Napoleon), it is important that you eat enough protein. Your metabolism breaks consumed proteins down into their amino acid building blocks which it can then recycle to make your own proteins, like muscle, hair and the enzymes that allow you to break down starch into sugars, among other things. These are what we call "essential amino acids". These are amino acids our bodies can't make, and we can only get them from the foods that we eat. Those celebrity chefs pushing paleo diets tell us we need to eat more protein in our diet, and avoid grains which are "unnatural". I can't put it better than the Paleo Way website:

> *"The Paleo Way has been designed to guide you as you transition into this wonderfully health benefitting lifestyle that draws on the core principals of our Hunter-Gatherer ancestors and modern-day knowledge and abilities."*

The Paleo Leap website tells us quite clearly that many foods are not allowed, because they didn't exist in the days of the caveman. Their advice states (*my bolding*):

> **"Cut out all cereal grains and legumes** *from your diet. This includes, but is not limited to, wheat, rye, barley, oats, corn, brown rice, soy, peanuts, kidney beans, pinto beans, navy beans and black-eyed peas."*

I didn't know these species did not exist back in caveman times, although it is true that many of them did not exist in the same form as they do now after domestication. But then, neither did chickens, cattle or sheep, which you are supposed to eat in abundance. Maybe we should bring back all that megafauna we hunted to extinction?

Now it's quite possible that you happen to be a human who can't eat too much meat because:

- You're vegetarian or vegan.
- It is precluded by your religion or beliefs.
- You just don't like the taste.
- You are trying to reduce your meat consumption for environmental reasons. Or
- It is too damned expensive.

Then perhaps the paleo diet is not for you. Same if you happen to be a pig. If you're a wild pig, then the hunter-gatherer lifestyle fits your situation perfectly. Just stop busting down those fences and eating sorghum and maize when it's ready to harvest – it's not allowed paleo-pig! If, however, you are, like most pigs, living on a farm with fences and walls, the paleo diet may be difficult for you to adhere to.

People who don't eat meat need to get their protein from somewhere else, and the most widely used form of plant protein is the legumes or pulse crops. Although it is also important to note that even at 10–12% protein, there is a decent amount of protein in wholegrain cereals as well. These plant-based protein forms can be pretty good too, except for a lack of a couple of essential amino acids. The cereals are typically deficient in lysine, and the pulses are most commonly deficient in methionine, which is a sulphur-containing amino acid. Animal nutritionists have been aware of these shortcomings for quite a long time, and free methionine (and sometime lysine and tryptophan) is typically added to most animal diets, given that those diets predominantly consist of cereals (wheat, maize, sorghum) and protein (usually soybean meal after the oil has been extracted). However, in the quest to make soybeans better for human and animal nutrition, one obvious solution was to find a genetic means to increase the methionine content.

So where can we find a good source of methionine? Brazil! Or more specifically, the Brazil nut (*Bertholletia excelsa*). Brazil nuts are high in oil. So high in oil you can light them with a match. I know this because one of my school friends demonstrated this after school one day. He got a lighter and lit a Brazil nut in the living room. When his fingers started burning he dropped the nut on the new carpet where it made a wonderful scorch mark. We were still attempting to clean the carpet when his mother came home from work. I was glad the experiment took place at his house and not mine. Brazil nuts are very high in methionine as well, and the main source of this was a specific seed storage protein known as the 2S albumin. Susan Altenbach from the USDA Center for Plant Gene Expression in Albany, California, led the research where it was first demonstrated they could introduce the 2S albumin gene to tobacco, where it was specifically expressed in the seed and resulted in a 30% increase in methionine content. They subsequently introduced the gene to soybean and canola, and other scientists at Gatersleben in Germany were taking the same approach for various other bean species. These new pulses opened up the possibility of redesigning the nutritional profile of grains to be more suitable for human and animal nutrition.

The use of the Brazil nut did ring a few alarm bells with some people. It had been known for some time that a certain proportion of the populace had an allergy to tree nuts. I should point out here that it was also known that a certain percentage of the populace have an allergy to soybeans. A known unknown was that it had not been determined which of the components of the Brazil nut was the cause of its allergenicity, however, there was pretty good evidence pointing at the protein component. What if the very protein transferred to these soybeans happened to be the Brazil nut allergen? What are the odds?

There are a number of ways in which you can test for allergenicity. Probably the most well known is the skin-prick test. This involves injecting a small amount of a solution containing the diluted allergen (sometimes in a dosage series) just under the skin of a group of people that includes those with a known allergic response, and those not known to suffer from the allergy. So in this case, it was important to select individuals with no known allergy to soybean, some of whom had a known allergy to Brazil nuts and some with no known allergic response. After 10 minutes a measurement is taken of what is called the wheal and flare diameter. The wheal is the raised blanched area and the flare is the redness surrounding the wheal. These measures give a semi-quantitative indication of the individual's reaction to the allergen. This research was performed by food scientists and clinicians at the Universities of Nebraska-Lincoln and Wisconsin-Madison, and was fully funded and disclosed by Pioneer Hi-Bred in the *New England Journal of Medicine* in 1996.[3] It demonstrated that the individuals with a Brazil nut allergy were also allergic to the GM soybeans, but not the wild-type soybeans. In addition, they took serum from susceptible individuals, and looked at the binding of proteins from the purified 2S albumin, and fractions collected from Brazil nuts, GM high methionine soybean and non-GM soybean to the IgE fraction of the serum. Clinical research confirmed that eight of the nine people with a Brazil nut allergy tested were also allergic to the purified 2S albumin and GM soybean expressing the 2S albumin.

Scientifically this was very exciting, because until these experiments, the Brazil nut allergen had been unknown. It took transgenic plants expressing a gene for a particular Brazil nut seed storage protein to give a clear indication of this. It also rang alarm bells. It was a demonstration that it was possible to transfer allergens and their subsequent allergies. A person who knew to avoid any foods with tree nuts could have, theoretically at least, now suffered from the same allergic response by eating these soybeans. Why only theoretically? Because as a result of this testing, the high methionine soybean was never released. The testing regime used to ensure that there were no problems with the GMO successfully identified a potential problem, and the decision was made not to proceed with commercialisation. There well may have been some internal arguments that perhaps the soybean could only be released for animal feed. However, Pioneer felt this was not a good option, as the consequences of the soybean getting into the human food chain were too great a risk to take.

However, this did not stop a media and activist storm. *The New York Times* led with the headline "Genetic engineering of crops can spread allergies" (14 March 1996, the same day the study was published). Many scientists stated that there was reasonable concern regarding the potential to create crops with different allergies, and this research showed that. What the research did not show was that GMOs = allergens, but that was the way it was painted by many in the activist world. And this continues today. The Australian Greens Party, who often hold the balance of power in the Senate state on their website:

- *Genetically modified organisms (GMOs), their products, and the chemicals used to manage them may pose significant risks to natural and agricultural ecosystems.*
- *GMOs have not been proven safe to human health.*

I am particularly fond of the last statement. Try it out yourself. Insert your favourite word at the beginning of the sentence.
[Insert word(s)] have not been proven safe to human health.
Select one of the following:

- Plants
- Tofu
- Rivers
- Bathrooms
- Toyotas®
- Semi-automatic weapons
- Kinder® Surprise
- Buttons
- Bees
- Organic foods
- GMOs

In fact, many of these are regularly involved in the death of humans, sometimes on a daily basis. Of this list, the only thing never actually demonstrably causing a human death has been GMOs. Secondly, how do you "prove" something is safe for human health? Actually it's impossible. All you can do is look at the associated risk. According to the CDC (who Zen Honeycutt doesn't rate as a plausible source), about 235 000 Americans suffer death or injury in a bathroom every year. How many suffer death or injury caused by GM plants turned into food? Zero. Both annually and cumulatively.

Once it had been demonstrated that these soybeans contained an allergen the project stopped. The pigs never got to eat the high methionine soybean meal, and this may well have been a lucky break for these pigs.

6.2 KINKI PIGS

To paraphrase George Orwell, not all pigs are created equal. The Kinki pigs were not so fortunate. OK, it's just me calling these animals the Kinki pigs, but they were the results of some particularly unclear scientific reasoning, and the outcomes were not what you might call good for the pigs. A group of scientists at Kinki University in Japan asked a question. One of those questions that kids ask their parents or vice versa when it comes to a dad joke.

The question they asked was, "What do you get if you cross a pig with a spinach"? While I am undoubtedly being flippant here, the outcomes of this research should have been foreseen and, perhaps in hindsight, the research should never have taken place.

Let me explain by starting with plants, animals and oils. Everyone knows that if you eat animal fat, you're gonna die. Everyone also knows that if you eat olive oil with every meal, you're going to live forever. Not sure what happens if you fry up your pork chops in olive oil? Do they cancel one another out? Maybe it's like celery, which has negative calories.[‡] Or maybe it just tastes like it has negative calories. But maybe if you add enough of it to your pork chop fried in olive oil you will have a really healthy meal. If you've got to the end of this paragraph and you're nodding, then I'm not coming to your place for dinner. So let's leave the celery myth out of the equation, and get back to those fats and oils.

Fact one – we need fats and oils in our diet, collectively known as lipids. Not too much, but they are definitely essential and they contain some of the key nutrients. Fact two – vitamins A and D are only soluble and hence only bio-available in the presence of some sort of lipid. That's one of the main reasons we're told to eat oily fish. All cells contain some fat, because the cell membrane is what is known as a phospholipid bilayer. Yet there is usually not enough lipid in cell walls for a healthy diet.

The major components of lipids are known chemically as fatty acids. The main types in animal fats (including our own) are stearic acid and palmitic acid, which are saturated fats.

[‡]Yes, I know, celery does not have "negative calories".

The main types in plants are linoleic, linolenic and oleic acid, which are unsaturated fats. Overall, the fatty acids are classified as saturated or unsaturated, with common classifications being:

- SFA – saturated fatty acids
- MUFA – monounsaturated fatty acids
- PUFA – polyunsaturated fatty acids

Then the acids are further classified with respect to how many carbon atoms there are in the chain of the molecule, and how many double bonds they have within the carbon chain. The double bond results in that fatty acid being unsaturated. A single double bond will result in a MUFA, whereas two or more double bonds result in a PUFA. Hence a fatty acid designated 18 : 0 has 18 carbons and no double bonds (SFA stearic acid), whereas a fatty acid designated 18 : 1 has one double bond (MUFA oleic acid) and 18 : 2 will have two double bonds (PUFA linoleic acid). Confused yet? Well there is another level of complexity, and if you are feeling lost, you need to go have some fish oil so you can increase your concentration. That's because that oily piece of mackerel contains omega-3 desaturated fatty acid. It is called omega-3 because the third carbon in the chain is the one that has the double bond (has been desaturated). These omega-3 fatty acids are the stuff that dreams are made of. Depending on who you talk to, or what social media you read, these oils prevent cancer, heart attacks, are important for brain development, help with mental health, and ensure you don't get dementia. But if you read other sources, there's not much evidence to support most of these claims, "but more research is required" as the Chief Investigator wrote to the Medical Research Funding Body or pharmaceutical company pleading for some research dollars/krona/pesos.

If you listen to all those nutritional experts and celebrity chefs, plants = good, animals = bad. As academic and food journalist Michael Pollan put it: "Eat Food. Not Too Much. Mostly Plants." The dogma is that animal fat (lipid) is saturated and bad while plant oil (lipid) is good. Especially if the darling wonder olive oil is extra-virgin cold-pressed on a dewy morning by doe-eyed 16-year-old extra-virgin nubile hand-maidens, or something like that. Or at least that's what Aldo wrote on the label, when he and his good for nothing brother actually did the pressing themselves in their tatty woollen jumpers, blue berets and a ciggy hanging out the corner of their mouths.

Back to the oils. Different plants and animals actually produce different ratios of these fatty acids, the reason being that some key enzymes can change fats from being saturated to unsaturated (mono or polyunsaturated), called fatty acid desaturases. The activities of these fatty acid desaturases will determine the fatty acid profile of particular oils. Not all olives produce the same oil profiles, just as not all peanuts do, or even all mackerel. It depends on the genotype, the environment, and in the case of the mackerel, what species it is, and what it was eating. Most nutritionists will tell you that olive oil is oleic acid, the wonder monounsaturated oil. While this is partially true, the oleic content of olive oil can be as low as 50% or can be over 80%. Linoleic and palmitic acid can be as high as 20% each, and there is often around 2% stearic acid (Table 6.1).

The same nutritionist will also tell you that animal fats are not healthy because they are unsaturated fats. This is partially true. Pig fat, for example, will tend to be mainly saturated fatty acids with around 21–25% palmitic and up to 25% stearic acid. However, pig fat will also contain significant proportions of unsaturated fatty acids, such as oleic, which may account for up to 40% of the fat, and linoleic, which is more like 5–10%. So that means that in some pigs, the oleic acid fat content is pretty close to the oleic acid fat content of some olive oils. In fact, palm and coconut oils are just as high in saturated fats as the pig, with palm oil usually about half palmitic (that's where the name comes from) acid, and coconut oil is highest in lauric, a $12:0$ saturated fatty acid.

Table 6.1 Fatty acid components of animal and plant lipids.

Type of fatty acid	$16:0$ Palmitic	$16:1$	$18:0$ Stearic	$18:1$ Linolenic	$18:2$ Linoleic	18.3
Pigs	21–25	2–3	10–12	43–46	9–15	<1
Cattle	18–26		10–19			
Canola	7	62		8	21	
Olive	14	71		<1	10	
Soybean	15	22		7	51	
Palm	49	40		<1	10	

As we all know, everything tastes better with bacon, right? Or else why would the Muslims go to such great lengths to make turkey bacon, or even more bizarrely why would vegans crave bacon so much that they have developed so many different products? The PETA website describes some of the best, like Phoney Baloney's Coconut Bacon, Sweet Earth Hickory and Sage Seitan Bacon or Tofurky Smokey Maple Bacon Marinated Tempeh. Vegan bacon .... It's definitely a thing.

For those of us who love the real thing, whether bacon, a grilled pork chop, a Bavarian bratwurst or a crispy skinned leg roast, there was help at hand. The scientists from Kinki University near Osaka had a plan to make pork a healthier meat. The addition of a plant gene, in this case, a fatty acid desaturase known as FAD2, was added to the pig genome. The thinking behind this was that this enzyme in spinach converts a MUFA to a PUFA, in this case, linoleic acid (18 : 2). And it worked, which was an interesting demonstration that a plant gene worked in an animal. The authors published their findings in *Proceedings of the National Academy of Sciences USA* in 2004. The transgenic animals had 20% more linoleic acid in their white adipose tissues than the non-transgenic animals. In the lipogenic adipocytes (the cells which make lipids), the level of linoleic acid increased from 2% to over 20%, and there was a significant reduction in the amount of 16 : 0 fatty acid (palmitic). Hence the pig fat became less like palm oil and more like linseed oil. The opinion piece published in the same journal[4] stated:

"This improvement could facilitate public acceptance of genetically modified food and pave the way for the commercial production of transgenic animals."

Interestingly, the paper also contained a few little statements like "The mortality of the F_1 piglets was relatively high", and in fact the numbers were a tad worrying. One of the surviving transgenic female pigs was mated with a normal male, produced 12 piglets, of which nine were stillborn. The three surviving piglets all died within three months, predominantly because the mother did not produce very much milk.

In the "what could possibly go wrong" category, have a little think about a pork chop. A pork chop has a nice (or not nice depending on your perspective) white band of fat around it, and we think of it as fat because it is solid. Of course, any chemist will tell you that it's a lipid that is solid at room temperature, which explains why when we heat that pork chop on the barbecue, a lot of the fat becomes liquid and drips onto the flames below. That's because stearic acid has a melting temperature of 69 °C, and palmitic acid has a melting temperature of 63 °C, so they are both solid at normal body temperature. This is most helpful, otherwise the fatty bits would get a bit sloshy in the animal body if they were liquid at 37 °C.

Now, imagine the main cooking oils in your kitchen cupboard. Firstly, we call them oils, not fats, because they are liquid at room temperature. Linoleic acid has a melting temperature of −5 °C, so it will remain liquid in the refrigerator. Oleic acid has a melting temperature of 13 °C, so if you put your olive oil in the refrigerator it should solidify. Try it at home. If it doesn't become solid, then you have a problem with your fridge (unlikely that it will be this warm) or your olive oil is not pure, or may even contain not a single millilitre of olive oil? NO! That would never happen would it?

Well, yes it has happened and continues to happen on a fairly regular basis. There is a lot of money involved in the olive oil business and the Italian "agro-Mafia" have been instrumental in making a lot of money by passing cheaper oils off as the more expensive super-duper mega ultra-cold-pressed extra-virgin olive oil from a small village in picturesque rural Italy. You may have heard and seen quite a few stories about this in the media over the past decade. Olive oil that was 70% sunflower oil. Extra-virgin olive oil that was just common low-quality olive oil with chlorophyll and beta-carotene added to give it the right colour, or Italian olive oil that could be traced to Spain, Greece, Syria or Tunisia. And consumers have been paying huge amounts of dollars for high-value oils, some of which was officially *lampete* quality. This means lamp oil from damaged or infected olives and deemed not fit for human consumption. An American living in Italy, Tom Mueller, wrote an excellent book, *Extra Virginity*, about the Italian olive oil industry in 2012, exposing many of the practices.

However, back to those pigs. Increasing the quantity of linoleic acid in pig fat (solid) is going to make that fat a lot more like a liquid (oil), which has to have health implications for the pig, and not good ones.

Interestingly, this was not the first time a GM pig project led to health issues. Some of the best documented problems came about after attempts were made to improve pig growth rates by expressing higher levels of porcine or other growth hormones in the animals. A USDA team in the 1980s added a gene encoding bovine growth hormone (bGH) and produced a number of transgenic pigs.[6] There were obvious signs of problems, with a significant number of the piglets born suffering from lethargy and lameness, as demonstrated by uncoordinated walking. In addition, many of the gilts (females) were anoestrus, and most of the males lacked libido. So when you have a bunch of females who are not producing fertile eggs, and a bunch of males who aren't particularly interested in sex, the survival of that particular genotype looks to be rather tenuous. However, like in all breeding experiments (whether animal or plant), there is genetic variation. Fortunately (or perhaps not) there were some females and males that were willing and able to reproduce, in at least some attempt to be the fittest who would survive.

After two generations of selection, pigs were produced that had some very positive traits. They had higher daily weight gain and improved feed conversion efficiency (which is how much feed does an animal need to eat to put on a kilogram in body mass). The best animals needed only 2.5 kg of feed to put on 1 kg of body mass, compared to the controls which required 3 kg of food. This was a big deal. The animals were growing faster (hence could be sent earlier to market), and required less food to do it (hence were cheaper to produce). So here was a very attractive productivity dividend. Another bonus, which was foreseen, was that once the animals were butchered, there was a significant reduction in subcutaneous fat. So here was a quality dividend, the ability to produce leaner pork, which was exactly what the consumer was looking for. Notwithstanding the many animals that did not do so well and were, to use the Darwinian parlance, selected against, this looked like a great use of the technology. That was until animal health and welfare aspects were taken into account.

The team measured some health and morphological parameters in five of the market weight pigs. Every single animal had gastric ulcers. Most had arthritis, dermatitis and other skin disorders, and cardiomegaly, which means an enlarged heart. Further analysis also showed that these animals had enlarged livers, kidneys and adrenal glands. None of the control (wild-type) animals showed any of these symptoms. These were all major animal welfare issues, and like the Kinki pigs with polyunsaturated body fat, these experiments never led to commercialisation of any transgenic pig. They did, however, also lead to an important body of information that was highly useful in understanding some of the nasty little "side-effects" seen in elite athletes, particularly from the Soviet bloc where the use of growth hormones and steroids appears to have become a standard part of training in the 1970s and 1980s.

In all the cases discussed above, whether the unfortunately allergenic soybeans or the supposedly "healthier and faster growing" pork, proper testing for any potential problems, as required by government regulations, had well and truly identified problems. None of these products ever made it even close to production. Hence none of these products ever made it close to being consumed in human foods. That was not the case with StarLink corn, as it was known in the USA.

6.3 STARLINK MAIZE

StarLink. What a fantastic name. If you wanted to elicit emotions about science gone too far into the realms of science fiction in our food supply, it would be hard to think of a better name, wouldn't it? Well, except for perhaps Darth Vader maize, or Voldemort corn. But Voldemort was still a product of J. K. Rowling's imagination, not having been published until 1997. StarLink corn became an almost overnight success as the anti-hero of the Frankenfoods anti-GM movement.

StarLink maize was the product of engineering a Bt gene to produce a protein that would break down in the plant or the gut of the insect less quickly. A more resilient insecticidal protein would surely lead to better resistance to the pest, and perhaps also would delay the ability of the pest to evolve resistance to the insecticidal protein. Scientists from a biotech company called Aventis (which came from Plant Genetic Systems, from the group at Ghent) decided that this was the Bt approach of the future.

The gene and protein were known as cry9C, and this was a synthetic version of a protein, engineered so it would survive a longer time in the insect gut. This was envisaged to ensure better and more durable control of insect pests. Aventis produced maize that expressed the cry9C Bt protein in all plant parts, and then licensed a seed company, Garst, a subsidiary of the Advanta group, to commercialise the seed in North America.

A protein that survives a longer time in the gut? There's nothing too new about that. There are many naturally occurring proteins that survive a longer time in the gut, being more resistant to the degradation by the actions of pH and enzyme activity. However, some of these highly stable proteins are known to be allergens. Aventis was asked by the FDA and EPA to demonstrate that the protein would not be allergenic. This is an expensive and lengthy process, and would have meant a delay in commercialisation by at least two years. Another option for the company was to make a new product wherein the Bt protein was not found in the mature grain. Surely they could have achieved sufficient protection against the corn borer by only expressing the gene in the vegetative tissues of the plant, including the husks that encapsulate and protect the developing grain? Although a scientifically attractive option, that would have taken several more years. Whether it was because Aventis were running out of money and needed to get a product to market as soon as possible, or whether there was pressure from shareholders to deliver or further funding would be withheld, or whether it was just corporate greed *via* expediency, perhaps we will never know. The approach taken was to then split StarLink into two separate applications, one for animal feed only and one for human consumption. In 1998, by which time Voldemort was already becoming more famous than StarLink, the EPA approved the commercial release of StarLink maize as an animal feed only. This turned out to be one of the worst decisions ever made by a government regulatory body when dealing with a GMO, and the repercussions were, and continue to be, far-reaching.

StarLink maize was released and was thought to be a good product to the farmers. It produced good yields and levels of resistance to the corn borer. In addition there was herbicide resistance in the form of a glufosinate detoxifying gene, hence agronomically it was a very good package. Theoretically, delivery to the grain storage and marketing specialists locally – or what US farmers call the elevator – shouldn't have been a problem, provided it was identified as a feed grain only. Theoretically.

Theory and practice can coincide and to a scientist, a molecular cuisine chef, or a football coach, it is indeed a beautiful thing. When theory and practice don't match, the consequences can be serious. And that is exactly what happened. Interestingly, the care factor appeared to be pretty close to zero, both on farm as well as during delivery of the crop to the elevator. The farmers were predominantly focussed on getting the best price. The traders were most interested in getting their orders filled, and, of course, also at the best price. Did anybody really care if some of these grains made it into the human food chain?

Yes. Somebody did care. Somebody actively went out and looked for evidence that StarLink maize had found its way into the human food chain. And they found it. Lots of times, in lots of places. They extracted DNA from maize-based foods, and tested for the presence of the cry9C gene in foods like tortillas, tacos, breakfast cereals, and other foods that are known to be mostly maize. Who was this "they"?

"They" were a coalition of anti-GMO activists called the Genetically Engineered Food Alert. This coalition included the Center for Food Safety, Friends of the Earth, and the Organic Consumers Association. The Organic Consumers Association started the March Against Monsanto campaign. The Center for Food Safety has been a major advocate for labelling GM foods. Friends of the Earth describe themselves as the "world's largest grassroots environmental network", and stemmed from the anti-nuclear campaigns of the 1960s. One of their most recent campaigns in Europe headlines with "It's time to close our countryside to genetically-modified crops for good".

The Genetically Engineered Food Alert sent samples from many maize-based foods, readily available in supermarkets and fast-food restaurants, off to a lab called Genetic ID to test for the presence of the cry9C gene. Just to be clear here, whole foods made from crop plants, as well as readily identifiable plants, like a lettuce or a tomato, all contain DNA. Each cell contains DNA in three compartments – the nucleus, where the chromosomes are, and the small organelles, the chloroplasts and the mitochondria. The DNA on the chromosomes will, much like the DNA in your cells (whether you are a human or a pig), encode around 30 000 genes. A GM plant may contain an extra 1–5 genes, depending on what it is. Genetic ID used the polymerase chain reaction (PCR), which is a very sensitive DNA amplification technique. PCR was famously thought up by Kary Mullis when driving from Berkeley to his cabin in Mendocino County, and completed in the cabin with the assistance of a bottle of red wine. Or at least that's the story he told in his Nobel acceptance speech.[5]

Next question on the who's who in this zoo? Who were Genetic ID? They were a company set up specifically for this purpose, and their Vice-President was Jeffrey Smith, the yogic flyer and swing-dance instructor. Genetic ID was set up in 1996 specifically to test for the presence of GMOs in food, and Jeffrey Smith became a Vice-President in 1998. The company now also does many other types of testing including food contamination, microbiology and gluten testing. The fact that they have been doing this for over 20 years is demonstration they are good at what they do. Two growth seasons after StarLink maize was released in the year 2000, the Genetically Engineered Food Alert group announced that they had detected the cry9C gene in Taco Bell taco shells. These taco shells were made under licence by Kraft, one of the world's largest food groups. I don't think it would be an understatement to say "and then all hell broke loose". Here was a food product containing GM maize that had not been officially cleared as safe for human consumption. It was supposed to be for animal feed only. And within a few months it was detected in many other different food products.

The Chief Operating Officer of Aventis at the time was John Wichtrich. Speaking at an AgBio Summit in Chapel Hill, North Carolina in 2016, he stated that this was the third biggest media story of 2000. That's a big call, because this was the year of the George W Bush *versus* Al Gore presidential election, which involved recounts in Florida and the involvement of the Supreme Court before announcing a winner five weeks after the election. But back to StarLink. Further testing demonstrated that over 300 foods and beverages had to be recalled as the cry9C gene was detected in them, including Budweiser, which is a beverage some Americans consider to vaguely approximate beer. You may not know this, but Budweiser is made with a special blend of rice, barley and maize. According to Wichtrich, Anheuser-Busch, the brewers of Budweiser, "tied up the sewer system in Philadelphia because of all the beer dumped".

Naturally, litigation and counter-litigation ensued. Aventis managed to avoid court, and negotiated settlements with government agencies (state and federal), seed companies, food processors, brewers, fast-food restaurants and even farmers. The effect of the StarLink controversy was so massive that maize prices actually dropped by around 7%. Farmers who had planted StarLink sued Advanta and Aventis. Farmers who did not plant StarLink took out a class action against Advanta and Aventis, who settled for $100 million. Overall, it is estimated that the total cost to Aventis was close to $1 billion.

One of the smaller settlements was $9 million, but this case was a big one in other respects. This settlement was with three people who claimed to have suffered allergic reactions after eating maize products known to have contained StarLink. In fact, there were over 50 independent reports to the FDA from people who had allergic reactions to maize products. This now became a public health issue. The CDC investigated 28 cases, and concluded that seven individuals displayed symptoms that were consistent with an allergic reaction. Like the testing described previously with the high methionine soybean, blood samples were taken from these individuals to test for any reaction with the cry9C protein. None of the blood samples had any reaction to the cry9C protein, so it could not be demonstrated that any of the patients had experienced an allergic reaction. Later work (paid for by Aventis at the request of EPA), showed that there was "not a high likelihood" to be an allergic response to the cry9C protein. Perhaps more compellingly, cooking and processing would mean that the protein would mostly be degraded, hence even if it was allergenic, the amount of cry9C protein would be many times smaller than what was needed to cause sensitivity. It is important to remember that the PCR technique is designed to detect the presence of cry9C DNA, but did not demonstrate that the protein was present/absent. Also important to remember is that the protein had not been demonstrated to be allergenic, but nor had it been demonstrated to be non-allergenic. As Buchini and Goldman[7] pointed out:

"Unfortunately, proteins that are allergens do not have properties that completely differentiate them from other proteins."

Developing allergic responses require, firstly, a sensitization, and then an allergic reaction. The sensitization appears to be dosage-dependent. Hence what is a hypoallergenic food in Europe may be the cause of significant allergy cases in Japan, as is the case with rice.[8] Allergies to kiwi fruit were never heard of until the 1980s when they became more widely available, and by 2015, 13 different kiwi fruit allergens had been characterised.[9]

To summarise the whole sad story, we had a potential allergen that was, in all likelihood, not an allergen. This protein, even if it was an allergen, was estimated to have been present in food at levels many times lower that what is required for sensitization. Yet the episode still cost Aventis somewhere between $500 million and $1 billion and involved massive recalls of food and beverages containing the gene, the cost of which may never be known. It also damaged the reputation of Aventis, Advanta and the various food companies, as well as heightened awareness that Budweiser contains maize.

Shortly after this all took place, in late 2001, the massive company Aventis decided to sell off its seed business and stick to their main pharmaceutical trade. They sold Aventis CropScience to the massive chemical company, Bayer, for $6.6 billion. Bayer then separated it from their core business as Bayer CropSciences.

In conclusion, as set out in these three case studies, there was a number of smoking guns. The general public, who in many cases had not given much thought to GM crops, was brought into the argument by the various sides (government, corporation, activist, scientist). To the general public, the issues of allergenicity and GM crops were now intertwined. Federal government agencies, especially the FDA and EPA, who were supposedly protecting the public and the environment from food contamination, had failed to do so on a massive scale. The EPA had made a critical error in licensing the new StarLink product for animal feed only, without overseeing or insisting on a monitoring and stewardship programme to ensure that the product did not enter the human food chain. Public perception was that the Aventis seed company had put profits before human food safety. Why had the company not waited until they had the data on the potential allergenicity of the cry9C protein? Why had the EPA not thought through the potential risks regarding the release of a GM crop for animal feed only, especially when the crop in question was maize, an important human food grain and human food ingredient in the USA?

> *"As we move through and beyond StarLink, we should not refrain from asking the hard questions and searching for better answers to the challenges raised by biotechnology. Some might argue that the StarLink episode will lead to greater government involvement...but it's important to remember that this problem may not have occurred had industry complied with the terms of its license."*
>
> Dan Glickman, USDA Secretary of Agriculture[10]

The GM papaya has improved farmer livelihoods in Hawaii, but still met with activist resistance. Eventually the technology prevailed, but activists were more organised in other places and ensured that it was not released in South East Asia. Banana growers see this example and are looking to the potential of using the technology to save their industry from the latest strain of Panama disease. Yet they are very wary of the consumer reaction to GM bananas, whether in Australia, Uganda or Germany. We have the case of GM soybeans that would have caused allergies in some people, and GM pigs that led to some serious animal health and welfare problems. However, the soybeans and pigs were rightly never commercialised or developed further. Self-regulation and adherence to proper testing ensured that they never made it to commercial production.

Finally we have the StarLink maize food recall, which was a combination of the effects of corporate lobbying and government mishandling. All of these scenarios have been used by anti-GM activists, and in the next chapter we will explore how the debate has played out, and will continue to play out. And like most games, there will always be some who play dirty.

REFERENCES

1. R. C. Ploetz, Black Sigatoka of Banana, *The Plant Health Instructor*, 2001, DOI: 10.1094/PHI-I-2001-0126-02.
2. S. N. Davidson, *Plant Physiol.*, 2008, **147**, 487–493.

3. J. A. Nordlee, S. L. Taylor, J. A. Townsend, L. A. Thomas and R. K. Bush, Identification of a Brazil-Nut Allergen in Transgenic Soybeans, *N. Engl. J. Med.*, 1996, **334**, 688–692.
4. H. Niemann, Transgenic pigs expressing plant genes, *Proc. Natl. Acad. Sci.*, 2004, **101**(19), 7211–7212.
5. K. B. Mullis, Nobel Lecture: The Polymerase Chain Reaction, *Nobelprize.org*, Nobel Media AB 2014. Web. 25 Oct 2017, http://www.nobelprize.org/nobel_prizes/chemistry/laureates/1993/mullis-lecture.html.
6. V. G. Pursel, C. A. Pinkert, K. F. Miller, D. J. Bolt, R. G. Campbell, R. D. Palmiter, R. L. Brinster and R. E. Hammer, Genetic engineering of livestock, *Science*, 1989, **244**(4910), 1281–1288.
7. L. Bucchini and L. R. Goldman, Starlink corn: a risk analysis, *Environ. Health Perspect.*, 2002, **110**(1), 5–13.
8. Z. Ikezawa, K. Miyakawa, H. Komatsu, C. Suga, J. Miyakawa, A. Sugiyama, T. Sasaki, H. Nakajima, Y. Hirai and Y. Suzuki, *A Probable Involvement of Rice Allergy in Severe Type of Atopic Dermatitis in Japan Acta Derm Venereol Suppl Stockh*, 1992, vol. 176, pp. 103–107.
9. A. Moreno-Alvarez, *et al. Children* 2015; 2(4): 424–438.
10. J. L. Fox, EPA re-evaluates StarLink license, *Nat. Biotechnol.*, 2001, **19**, 11.

CHAPTER 7

Paint It Black

"If they think this is the way to go ... we [will] end up with millions of small farmers all over the world being driven off their land into unsustainable, unmanageable, degraded and dysfunctional conurbations of unmentionable awfulness."
Prince Charles, Heir to the Throne of the
United Kingdom, 2008[1]
(and Australia, Jamaica, Cook Islands, Canada,
New Zealand, the Falkland Islands, Cayman Islands,
Bermuda, Isle of Man, Turks and Caicos, ...)

Prince Charles saw a red door and he wanted it painted black. "Black as night, black as coal, I want to see the sun blotted out from the sky." No, that was Mick and Keith from the Rolling Stones, but it certainly sounds like a doomsday version of the future being painted by the person who is going to be my next Head of State (and yours depending on where you live). He's going to be on our banknotes and coins. Around 20 years ago, perhaps bored with attacking the evils of modernist architecture, the Prince of Wales became the darling of the anti-GM activists, especially in the UK, when he effectively became an anti-GM activist himself. The first "red door" that anti-GM sentiment really started to attack was the very first GM plant to be

Good Enough to Eat? Next Generation GM Crops
By Ian D. Godwin
© Ian D. Godwin 2019
Published by the Royal Society of Chemistry, www.rsc.org

commercialised, the Flavr Savr tomato. They wanted the red of Flavr Savr tomatoes "painted black", by whatever means it took. Talk about "unmentionable awfulness".

In 1994, supermarkets in North America began selling Flavr Savr tomatoes. In 1996, supermarkets in the UK started selling small cans of processed tomato paste. These cans were clearly marked as being from Flavr Savr tomatoes and were heralded as the next big thing in foods. And in Sainsbury's and Safeway they were cheaper than the others, from which no noticeable effort had been made to save any flavour whatsoever. Well at least that was not stated on the label!

Firstly, what was a Flavr Savr tomato? It was a tomato genetically engineered to have slower ripening, hence harder fruit for a longer time. This meant it was easier to pack and transport without damage when harvested. If you've ever grown tomatoes, you will know that green tomatoes are pretty hard, and red tomatoes are considerably softer. As the tomato ripens, the cell walls get broken down. A few other things occur that make the fruit generally more attractive as well – the fruit starts to change colour, heading towards that really beautiful tomato red that is predominantly caused by the biosynthesis and accumulation of lycopene (Figure 7.1). The fruit also becomes less acidic, and sweeter as more sugars accumulate in the fruit. If you leave it for too long, it gets so soft that the skin breaks, fungal and bacterial organisms cause the fruit to rot, and it doesn't smell so good, either. So you take it to the cinema and throw it at the screen if you don't like the movie, although the rotten tomato thing is predominantly now done in the figurative online sense. More civilised but perhaps not nearly as gratifying.

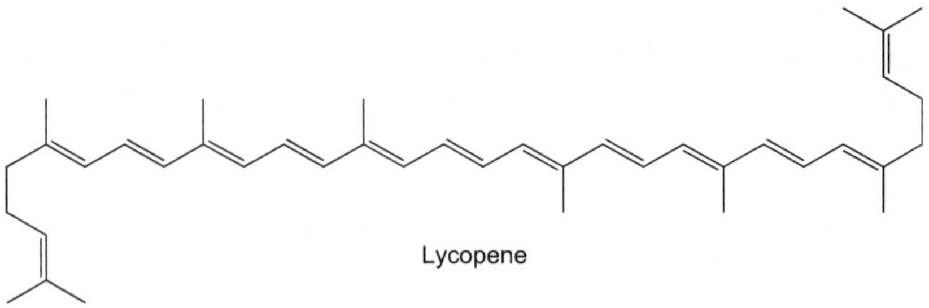

Lycopene

Figure 7.1   Structure of Lycopene.

The main component of the cell wall that keeps it rigid is pectin. If you are a jam-maker you will certainly have heard of pectin. Some of my earliest food science lessons with my mum were soaking citrus seeds in warm water to release the pectin, so it could be added to the marmalade she was making to ensure it was of the right consistency. And if you use over-ripe fruit, like my dad tends to when he makes his favourite strawberry jam, there is not enough pectin. Hence you need to add pectin (which you can buy as a powder) or you will get runny jam. My dad calls his concoction strawberry sauce to have over ice-cream, and it is excellent for that purpose. Why does ripe or over-ripe fruit have less pectin? As the fruit ripens, there is an increase in an enzyme called polygalacturonase (PG), and this is the enzyme that breaks down the pectin in the cell walls. Hence the riper the fruit, the softer the cell walls. For most fruits, we prefer them not to be rock hard. Biologically it's also crucial if the seed is going to manage to make it out of the fruit to start a new generation.

Transporting ripe tomatoes around is a tricky business. Even if you have an organic farm and you hand-pick the fruit on moonlit mornings and take them straight off to the local farmers' market on your bicycle, you have to be extremely careful. You need to avoid potholes, and things such as cobbled streets. Strap a lunchbox of ripe tomatoes to the back of a bike in the Paris-Roubaix ("Hell on Wheels") race, and you would only have to add some basil, garlic and cucumber when you get to Roubaix, and you'll have some delicious gazpacho – or maybe vodka, Worcestershire sauce, salt and a celery stick, because after that ride you deserve a better reward than cold tomato soup. And remember, celery has negative calories.[†] You get the picture. This is a reason why many tomatoes are picked green, and then treated with ethylene in a shed, which is a ripening phyto-hormone in gaseous form, so they can ripen before going to the local supermarket/farmer's market/ketchup factory.

[†]Please do not write to me about negative calories. This is one of those food myths: supposedly, you use more calories digesting these watery, fibre-filled extended plant petioles than they contain. In fact, they contain about 12 times the caloric content as is required to digest them, so it is rubbish. Yet there are many helpful dietary advice sites out there repeating this nonsense claim. Mind you, there are claims that are bigger nonsense, some of which are quite dangerous. Like GM Flavr Savr tomatoes will kill you, and the environment.

If you read the US literature, there was only one Flavr Savr tomato with delayed fruit softening. And softening is the correct term, because other components of ripening were to continue at the normal pace (like going red, accumulating sugars, reducing acid content). This work was led by a California biotech company known as Calgene. Calgene had famously started up in 1980, in a garage in Davis, California, as a joint venture between University of California, Davis Professor Ray Valentine and a budding venture capitalist, Norm Goldfarb, who was then working at Intel. Calgene was something of a biotech wonder in the 1980s, going from strength to strength, having first developed a means for herbicide tolerance which was licensed to Monsanto. However, their Flavr Savr technology became the first GM crop plant for commercialisation, released in the USA in 1994. Soon tomatoes with "longer shelf life" were available to consumers in their local grocer and supermarket. The tomatoes were not a screaming success for a couple of reasons. Calgene was not a plant breeding company and did not ensure they were creating their new tomatoes in the most elite germplasm. Hence they were not the highest yielding or tastiest varieties, and therefore not preferred by either the farmer growing these tomatoes, or the consumer indulging in them.

The approach taken on the other side of the Atlantic was a little different, and the development of tomatoes that softened less when ripening was something of a transatlantic battle. The team that worked out the nature of fruit softening and the degradation of pectin was actually led by Don Grierson, a Professor at the University of Nottingham. In the 1980s his team were able to pull apart many of the processes involved in fruit ripening, by using a technique then known as antisense to silence some of these genes. When I was a young postdoc in Birmingham, we had a lab meeting discussion about their recent *Nature* publication,[2] in which they demonstrated that downregulating the expression of PG led to delayed fruit softening in tomatoes. Two years later they followed up by showing they could downregulate ethylene production in tomato fruits, and both techniques could be used to improve the shelf life of tomatoes.[3] Much of this work was performed in collaboration with ICI Seeds (who became Zeneca in 1993, who slowly sold off most of their seeds business, much of which became Garst who shortly after merged with Dutch company VanderHave. The merged entity formed the company Advanta as discussed in Chapter 4).

The delayed ripening tomato therefore took two somewhat different paths. In North America, the Flavr Savr technology was released as fresh tomatoes. The Nottingham/ICI version was not released as a fresh product, but became available as canned tomato puree in 1996. It was released in supermarkets such as Sainsbury's and Safeway, selling for the same price as the conventional tomatoes (29 pence), except you got a 170 g can rather than a 142 g can. That is because the cost of production was about 20% lower with the new GM tomatoes. Furthermore, the product was clearly labelled and sold as a GM product. As Grierson said to Alison Goddard in an interview for *The Times Higher Education Supplement* (published in 1998 as an article entitled "A puree genius at his work". See what us plant scientists have to put up with):

"The key thing is that the product is labelled. This is not a scientific issue. It is not a safety issue. It is just a question of recognizing people's individuality, and their concerns, and their right to know."

Very prescient words. It's part of our scientific endeavour that we have debates with colleagues, both public and private, regarding our research. These debates can be well mannered or otherwise. Scientists get trained to do this from their undergraduate days, and also develop skills in showing governments, research funding bodies and corporations why our science has importance and positive impact. What we mostly had not been trained to do was advocate for our scientific advances with the general public. Grierson went on to say this was essential:

"by talking, explaining, by learning to explain at an appropriate level for the people who are listening. And, I have to say, by getting furiously mauled by people in the media who want to take a particular line."

Which suggests that the mauling may have started. And the outcome of the mauling was that the Flavr Savr tomatoes had only a brief few years in consumer shopping baskets before they disappeared. The wheels fell off.

When the product was launched, a BBC article (5 February 1996) stated that the new tomatoes had "the rotting gene removed". According to Zeneca, the tomato was a win–win for the farmer, the consumer and the environment, and in addition the tomato paste would have a stronger taste and "sticks better to pasta". For the newly minted "Frankenfoods" anti-GM groups, it was a call to arms. They promised a consumer boycott. Along with my plant scientist colleagues we were stunned, but the general feeling was that all we had to do was explain the technology and what we had been doing and most people would understand that this was safe and the way of the future. We were wrong.

The boycott and the Frankenfoods movement had a big win. Considerable sections of the British population were convinced GM foods were bad for health, bad for the environment and morally wrong. Just over three years later, Sainsbury's announced that it was going to phase out all GM food from its stores, including the canned Flavr Savr tomatoes.

"Our customers have indicated to us very clearly that they do not want genetically-modified ingredients in their food and we are taking steps to offer that guarantee."

Sainsbury's spokesperson, BBC article, 17 March 1999.

Sainsbury's had joined a European consortium of supermarkets who had decided that either: (i) there was no market advantage from stocking any GM products, or (ii) there was a market advantage in not stocking GM products and proudly declaring that this was their policy. Other British supermarkets were joining this trend, as were large supermarket chains in France, Italy, Ireland, Belgium and Switzerland. How did this happen?

Klaus Ammann is an Emeritus Professor from the University of Bern, formerly the Director of the University Botanical Garden until he retired on 2006. Like many of the activists, Klaus was very much aligned with the anti-GMO brigade in the 1980s. By nature left-wing and liberal, and perhaps something of an aging hippy, it seemed that what most of his friends were saying was right. By his own description, Klaus was into biosystematics (or what we once called taxonomy), and a "typical green". He had contributed to a book on the flora of China, and felt that plant conservation was a big challenge for agriculture. To him and his friends, GM crops and food were bad for the environment, posed all sorts of health risks, and most tellingly were an instrument of evil in the hands of large multinational companies, but mostly Monsanto, and other chemical companies that gave the world Agent Orange.

I had met Klaus at a number of conferences, and my first impression still sticks. This man is Santa Claus (Father Christmas, St Nicholas, Julemanden or whoever is your preferred Christmas visitor and bringer-of-gifts). He is jolly, has beautiful red cheeks and eyes that twinkle through his wire-framed glasses, a flowing white beard, and he seems to have a natural predisposition towards laughter. When I was fortunate enough to interview him, in Neuchatel where he is currently living, none of that had changed.

Klaus told me he was a fairly early convert to being a true believer in the value and potential of GM crops, and it came after a meeting with Ingo Potrykus, one of the pioneers of GM, and a co-inventor of Golden Rice. At the time, Ingo was a group leader at the Friedrich Miescher Institute in Basel, and was (as described in Chapter 2) a key player in the early development of GM plants. Ingo invited Klaus to come to his lab for a week, and see what was involved in making GM plants, and get an understanding of the technology and how it could be used to benefit human health and the environment. After a week in the Potrykus lab, Klaus came fairly rapidly to the conclusion that he had been wrong. Not only did he then go and tell his friends about it, he became so incensed with the quality and veracity of the public debate that he changed his public life and research focus totally. Klaus became one of the fiercest pro-GM scientists in Europe, and now in his mid-70s he shows little or no sign of slowing down. He still has so much to say. In fact, he always has had a lot to say. Klaus is one of those hyper-enthusiasts who will fill a 20-minute conference slot with enough slides to last the average lecturer a 2- to 3-hour lecture slot. Undeterred, Klaus will power his way through a rollercoaster of an entertaining presentation. When you get to the end, you are partly glad it's over, but also wish you could experience it again because your senses tell you that there was a lot that you missed. That was exactly how I felt when I finished talking with him. He has the ability to say exactly what he is thinking. It comes out with such alacrity and conviction that one has to question whether the thought was fully processed in his brain before his utterances. So much so that I had to call him back for clarification of a few issues, and make sure that I hadn't misunderstood or taken him out of context. He felt that typical techniques of fascism had been used by the Greens to stigmatise GM crops and foods. He feels that the term "genes" has replaced the term "Jews". The quote below is one of the more provocative illustrations:

> *"It worked for the Nazis when they made the Jews the bad guys. They worked to stigmatize the Jews. The anti-GM activists, the Greens and Greenpeace, used exactly the same methods to turn the German public against GM crops. They didn't use facts, it was all just mind-games and propaganda, and now their minds are closed to further debate!"*

Switzerland, like many countries in Europe, imposed a moratorium on the cultivation of GM crops. Not being a member of the EU, Switzerland went to the people, who voted in favour of a five-year moratorium in 2005 (55.7% in favour of the ban). This was extended by parliament in 2010 and again in 2013, such that the moratorium was reviewed and extended in 2016 until 2021. When I spoke to Klaus in late 2016, he was optimistic that after this time, there would be some cultivation of GM crops in Switzerland.

He has always advocated that "every new technology needs to be regulated, but in a dynamic and scalable way". Regulators are afraid of doing the wrong thing, as are politicians. Klaus has served on the biosafety committee in Switzerland for eight years, and feels that there are many politicians who are now confident that GM crops are largely safe. The problem is whether their place as a representative of the people is safe if they voice this opinion publicly. There is the added problem of training and attracting young scientists. As he pointed out, young scientists in Switzerland and Germany were either shying away from GM research, or leaving the country to work in North America, where they felt their work would have impact. As we will discuss in the next chapter, the same phenomenon is having far-reaching impact on Swiss and German companies.

In 2012, I visited collaborators at the University of Zurich. We had been demonstrating that a key leaf rust (caused by *Puccinia triticina*) resistance gene from wheat (*Lr34* originally cloned by Evans Lagudah at CSIRO) conferred not only rust resistance when transferred to sorghum, but also resistance to the disease anthracnose, cause by a totally unrelated fungus, *Colletotrichum*.[4] Just for total clarity, when I say "transferred to sorghum", I mean using GM techniques because there is currently no other way to transfer a gene from wheat to sorghum. My collaborator, Simon Krattinger, who has now taken up a well-funded Professorship at the King Abdullah University of Science and Technology in Saudi Arabia, took me to the Agroscope field station at Reckenholz, north of Zurich. We were met there by Jörg Romeis, the Head of Biosafety at Agroscope, which is the Swiss centre of excellence for agricultural research. Jörg and Simon then took me out into the field where we then drove beside a 3 metre high fence with razor wire on the top. Whilst driving along a large German Shepherd dog ran along parallel to the car on the other side of the fence. We then stopped at some gates and were met by a security guard who let us in through one set of gates, and then a second gate. It became apparent that the fence was actually electrified, and there were two further fences inside the main perimeter. Inside the inner gate was a mobile home, which served as a 24/7 accommodation for the guards, and of course the dog. This establishment is one of the few places in Switzerland where GM crop field trials are allowed to take place.

Beside the accommodation block was a tower with a number of cameras mounted about 4 metres above ground level. When I asked about the cameras, Jörg explained they are not just ordinary cameras but also night vision and infrared. Apparently on a number of occasions, anti-GM activists have massed in the forest on the perimeter of the enclosure, hoping to storm the barricades before a defence can be mounted. So long as there is an infrared camera, it can pick up movements inside the forest, and give warning to security and scientific personnel. It has happened a few times, and most often it seems to have been a cow who has wandered into the forest for a bit of arboreal exploration. While I was there it was very peaceful and pastoral. We walked around and looked at the GM trials of potatoes, wheat, canola and tobacco, and the only activity was the guard dog chasing a ball that the security guy kept flinging – probably the only way they can both prevent the risk of dying from boredom and obesity. We had a rather dreary and pessimistic conversation regarding organic potatoes, diseases and copper (more later), and then went back to the main buildings, where we discussed anti-GM activists.

Naturally, I had seen some of the images of these activists destroying GM trials in various parts of Europe, but I had not, until that visit, appreciated the full scale of the destruction across Europe. I had also not appreciated that these trials were not solely the activity of large seed companies. Many of them were in the public sphere as well, and a significant proportion had been set up for the specific purpose of studying biosafety aspects, such as pollen flow and other perceived or real effects on the environment. Marcel Kuntz, a government scientist working with the Institute National de la Recherche Agronomique (INRA) and the Centre National de la Recherche Scientifique (CNRS) in Grenoble, France, documented some of these.[5] In all, he reported on 80 acts of vandalism to government or public institution GM field trials across France, Germany, the UK, Switzerland, Italy and Belgium. In Germany, there was a large number of events, many involving Justus Liebig University, Giessen and their maize and barley trials. Similarly in France, there was vandalism of canola, wheat, maize, grape and coffee research. The coffee trials were being performed by the French international development agency (CIRAD) in Montpellier. In England, there had been 256 plantings of GM crops since 2000, and 41 of these had been vandalised.

From the outset the anti-GM activists often worked with the media, to ensure that their message was seen by many. They liked to portray their anti-GM credentials as being a grassroots movement. However, they really needed to ensure they used media and later, social media, to attempt to get their message to the grassroots. When activists were then taken to court, and fined or sent to gaol, the movement, by necessity, became more underground. Special forces-type raids in the middle of the night continued, but the public destruction became less frequent.

Marcel Kuntz talked about the shift in the attitudes and the means of operation of these groups, some of which were affiliated with larger groupings like Friends of the Earth or Greenpeace, but increasingly starting their own small groups. Scientists also started to take action themselves. Not only were their beliefs and own livelihoods being challenged, they were becoming more and more incensed that the general public seemed to be increasingly gaining a mindset against GM crops and foods. Instances of confrontation became more widespread. Kuntz described one such instance in 2012:[5]

"In Belgium, at Wetteren, on Sunday May 29, 2011, a movement calling themself the "Belgian Field Liberation Movement," assisted by French activists, damaged an experimental field of GM potato resistant to late blight (caused by Phytophthora infestans) which had been implemented by a consortium consisting of the University of Gent, ILVO, VIB (Flemish Institute for Biotechnology) and the University College of Gent (http://www.ogms.be/actualites/wetteren-135-000-euros-de-dommages)*. Consortium scientists had attempted to dialog with the activists on their 7th of May "gene spotting" activity. A broad group of scientists then organized themselves, using the motto "Save Our Science" in an attempt to call upon the protection of this field research. On May 29th about 300 people gathered to protect the field trial site, while about 350 people joined the opponents' meeting. A police force of 86 was present. The activists agreed that around 30 people would try to destroy the trial without violence and would offer no resistance to arrest. But after the arrest of some individuals, almost all the others invaded the trial premises. A number of police officers got wounded in the subsequent violence. Finally, 18 people succeeded in getting through the fences and 7 of these were able to reach the potatoes and destroy 15% of the trial."*

In Germany, activists were able to get prior information as to where field trials were to be planted. One of their new tactics was "occupation". They would occupy the field, often building an observation tower, and sometimes chaining themselves to the tower. One of Germany's largest seed companies, KWS, is based in Einbeck in northern Germany. KWS is a market leader in sugar beet varieties and has a considerable market in oilseed rape and other crops. In 2008 and 2009, a number of field trials of GM sugar beet were destroyed by activists. Other fields were occupied by activists with a new trick – some of the protestors trapped their hands in barrels that had been filled with concrete. At Northeim in 2008, 450 KWS employees joined hands around a field to keep protestors at bay while a GM sugar beet field trial was planted. The Northeim field site was not very far from the KWS SAAT headquarters and research laboratories just outside Einbeck. This is not so easy to achieve when the field trial is in a more remote place.

My colleague and friend, Professor Jimmy Botella, is a plant biotechnologist at the University of Queensland. Jimmy has been in Australia since the early 1990s and has had his share of interactions with anti-GM activists, particularly when his group were undertaking field trials of GM papaya and pineapples. He explained to me that one of the greatest production limitations is the lack of synchronisation of flowering. For pineapple production, farmers use a product called ethephon, which is the most widely used plant growth regulator (phytohormone) in agriculture. Once it is sprayed onto the plant it is rapidly converted into ethylene, the ripening/maturity phytohormone, which is also used to trigger ripening in fruits such as bananas and tomatoes. Pineapples are notoriously poor in their regulation of flowering and if they flower too early or too late, this results in smaller fruit. Smaller fruit get lower prices at the fresh fruit market, and are effectively not marketable for canning. Can you imagine the consumer complaints if the pineapple rings inside a can were only half the diameter of the can? Another alternative is for the farmer to harvest at different times. This is where the economics don't work. Even with the single harvest window, labour-intensive harvesting is the most expensive part of production, making up over one-third of the total cost of production.

In the 1990s, over 70% of the cultivated pineapples worldwide were a variety known as Smooth Cayenne (smooth because the leaves did not have any spikes). Smooth Cayenne, while not the sweetest and most fragrant pineapple, was perfect for processing and canning. It is just the right size to peel and process and slot into the can (with added sugar). Jimmy's GM pineapples used gene silencing technology (RNAi) to downregulate a gene involved in ethylene biosynthesis in the plant. By downregulating a gene known as ACC synthase, the plants would no longer be able to make their own ethylene, hence they would never flower without the exogenous application of ethephon. This overcame the problem of early flowering, which in many years was a problem with at least 30% of the plants in a single paddock. When 30% of your plants don't produce harvestable fruit, the economic impact is not just serious but will usually represent the difference between profit and debt.

In 2000, the GM pineapples were planted on a site near picturesque Redland Bay, on the coast near Brisbane and away from the pineapple production areas. One day, when Jimmy was at the site, a small group of activists held a protest outside the facility. Then a couple of days later, he received a phone call from the police, telling him that activists had broken into the site and were in the process of uprooting all his pineapple plants, and while this was in progress all the local TV channel news crews were filming it. This is in a place where the likelihood that the TV cameras (all four local TV stations) happened to be driving past was quite minimal. Jimmy hurried to the scene. His recollection is that there were eight activists in total. These eight activists had uprooted all of his pineapple plants and were talking to cameras to denounce the environmental evils and pollution of GM plants. When he watched the news that night, he was astonished to find that the TV stations had somehow managed to make it look like there were 200 protestors or more. The next day he received a note from the "Seed Liberation Army" outlining their manifesto to destroy any GM plants and save all the "natural" seeds for future generations of farmers.

After the initial shock of events had diminished, Jimmy took most of it with his characteristic good humour. He found two things very funny. Firstly, pineapples don't have seeds and have to be reproduced vegetatively by cuttings known as slips. Hence like bananas, pineapple production is performed by growing a large clonal population. So despite the best efforts of the activists, not a single seed was saved or destroyed. Secondly, pineapples are bromeliads and use a form of photosynthesis known as crassulacean acid metabolism (commonly called CAM photosynthesis). CAM photosynthesis is an evolutionary development favoured by desert plants, such as aloe, agave (used to make tequila) and most cacti. It means these CAM plants can survive long periods without much water being available. A key part of their adaptation is that they close their stomates during the day, and hence the tough leaves lose hardly any water during the heat of the day. They then open their stomates and continue photosynthesis late at night using the light energy they have chemically stored during the day. As a result of CAM photosynthesis, Jimmy and his team were able to place the plants back in their positions, apply irrigation, and total losses were only three deaths out of almost 800 plants. The plants were slowed down in their development, because they had lost most of their roots, but they soon grew them back. You can observe this phenomenon at home (yes, you can try this at home) by taking a pineapple top and putting it in the ground. The top will soon form roots and you will have a pineapple plant. I once had a puppy who decided the ultimate fun was to pull the bromeliad plants out of my garden and run around with them in her mouth while I chased her. Never lost a single plant if I replanted it within 24 hours. There was that weekend I went away, however, and the garden was cured of bromeliads.

It may even be that because the GM pineapples were unable to make their own ethylene (the senescence hormone) their survival rate was higher too. So the major outcome of the activist activity in this case was to give Jimmy a funny story to tell over a beer. Well, that and the fact that he had to fill out a lot of incident reports!

Many of the activists were driven by the desire to protect the environment and, in some cases, human health. This explains why many of the first high-profile anti-GM activists affiliated themselves with either the eco-warriors of Greenpeace, Friends of the Earth or the Worldwide Fund for Nature (WWF). Many of them have had little or no personal experience with the production of food. Why were anti-GM activists so strident in their opinions? Why did they feel they had the moral high ground and that they could wilfully damage experiments, many of which were set up to generate information regarding likely effects on the environment? Experiments that were aimed at measuring:

- The evolution of weediness.
- The evolution of pest and insect resistance.

- Effects on non-target insect populations.
- Pollen (and hence gene) flow to other crops or wild relatives.
- Impacts on soil health and micro-biodiversity.
- Nutritional quality when compared to conventional genotypes.
- Nutrient use and water use efficiency.

Surely these were the very experiments necessary to answer many of the questions that environmental and health activists had been asking? The plant science community was literally trying to generate the appropriate data, yet our efforts to do so were being chopped down, uprooted, destroyed with brush-cutters, or poisoned. The poisoning of GM field trials with toxic chemicals or herbicides was one of the strangest phenomena to plant scientists. GM plants were being created to reduce dependency on insecticides and fungicides, yet some activists were using pesticides to destroy the research. This was a mind-boggling double standard.

However, given the sheer number of GM field trials, and the commercial cultivation of GM crops on a broad scale in the Americas since 1996, within a decade there was a great deal of data generated to counter many of the claims that had been made based on environmental concerns. As discussed in previous chapters, in most cases, there were socio-economic and environmental benefits in places where GM crops were grown. In cases where the touted benefits were not experienced, farmers soon reverted to their previous favourite cultivars and managements systems. If you want to try something new and it provides no obvious benefit, revert to what you did before, or try something else new. In the face of scientific data, the very credibility of some anti-GM groups was called into question. And that's when, according to Marcel Kuntz from the National Centre for Scientific Research, the problem for the credibility of anti-GM groups began:[5]

"Whilst environmental groups initially raised sensible concerns about potential environmental damage, they soon shifted to an ideological position of opposition, as science demonstrated that such risks are often small, sometimes hypothetical and generally not specific to GMOs. Given the lack of scientific evidence to support the purported health or environmental effects of GMOs,

opponents have moved on to attack the risk assessment of GM crops. Scientific authorities are not only questioned on the quality and honesty of their experts—which is unpleasant for them but a matter of legitimate debate—but also attacked, by postmodernism, on the scientific method and its universality."

There are quite a few issues that keep being recycled in this debate, and they are just not correct, or nowhere near the full truth. Some of these arguments that have been used, recycled, repackaged and reformatted include:

- GM foods are "toxic".
- Over 400 000 farmers in India have committed suicide when their GM Bt cotton crops failed.
- Bt corn killed Monarch butterflies.
- Farmers are "forced" to grow GM crops.
- All GM crops require toxic chemicals to ensure they can grow, and hence are laced with toxic chemicals, ie "GMO = pesticides".
- GM is not organic.

There are many, many more, but I am now going to discuss these without getting angry or opening a bottle of red wine (I didn't say anything about beer). However, a strong driver in the face of activist faith in the horrors of GMOs is one of the bases of our human nature. As Joshua Green wrote in his book *Moral Tribes* (2013),[6] we evolved under tribal conditions and still maintain tribalism in many ways:

"Our brains are designed for tribal life, for getting along with a select group of others (US) and fighting off everyone else (THEM)."

We all tend to identify with people who have similar "values" as us. And worse still, many of the values tend to divide us in a binary manner, when the reality is always more complex. Do the test on yourself:

- Brexit: Leave or Remain?
- Trump: Making America Great or Dangerous Megalomaniac?

- Dogs or Cats?
- Left or Right?
- Vaccines: Good or Bad?
- Fox or CNN?

And then it comes to football teams. Here we know it is complex, but we tend to lump the supporters of other teams all together as THEM, and only our supporters (US) are the best people. And that is something ridiculous and every year the arguments result in deaths. Really stupid, pointless deaths.

I have had collaborative research projects with Papua New Guinea for over 20 years. One of the biggest events on the calendar is the State of Origin Rugby League series. It's not really a good time to go to Papua New Guinea, and if you do, stay home, steer clear of windows, and don't wear anything maroon or blue. Unless you're in a place where everyone else is wearing one of these colours, in which case opt to conform. PNG is probably the only country in the world where rugby league is the national sport. The State of Origin is when Queensland plays New South Wales (these are Australian states, not part of PNG) in a three-game series. "State against state, mate against mate". This is such a big deal in Papua New Guinea that every year there are violent altercations, often resulting in death, over perceived bad refereeing decisions or on-field violence spreading through live television. In 2009, three men were killed in separate violent altercations, and similar events occurred every year. In 2015, after three deaths associated with arguments resulting from game 2, police called on the government to ban showing the next game on live TV. This is truly vicarious tribalism. Do the Danes get violent when watching the German Cup Final? Do they even care? Do Scots hurl beer glasses at each other when they watch two English football teams? (Actually, many of the Scots I know are probably wishing both teams will somehow find a way to lose.)

7.1 GM FOODS ARE TOXIC

GM foods are not toxic, and as already mentioned, not a single person has died from eating GM crops that had anything to do with the GM trait. People have a general perception that "organic" foods are safe whereas they may be quite a bit more circumspect when you use the words "GMO". Yet there are well-documented cases where people have died as a result of eating organic food. However, I will discuss organic agriculture and food in detail in Chapter 8, so we will get back to the "toxicity" of GM foods. It is easy to find "information" about this, you can ask Dr Google.

I just asked (search: GM food is toxic), and Dr Google provided me with lots of great hits. The top hits that were not government or university websites (saying the toxicity of GM food is a myth and then debunking the myth) were:

1. Hang the Bankers (an alt-right propaganda, conspiracy theory website).
2. Earthopensource (run by John Fagan who is a Professor of Microbiology at the Maharishi University of Management).[‡]
3. Greenpeace (no introduction needed).

In 2002, Zambia was faced with famine. Bad weather in the form of floods in 2001, followed by drought in 2002, caused severe shortages of the main staple food, maize, which was normally milled and eaten as meal (porridge). By August 2002, over 2 million Zambians were suffering from food deprivation. Donors, through the World Food Programme, were supplying maize. A significant proportion of this maize was from USAID, and Greenpeace moved in to prevent 35 000 tonnes of maize from being distributed to starving people. Greenpeace convinced the government that the maize was GM, and was therefore toxic, leading to the Zambian President, Levy Mwanawasa, calling the maize "poison". One of Zambia's major concerns was that farmers may keep the seed and grow it, and that the GM seed would have jeopardised the Zambian trade of maize to Europe.

OK, breathe in. Don't open that beer yet (I'm talking to myself here – if it's beer o'clock where you are, you do as you see fit). While it is true that a farmer may have kept the seed to plant, they would have been sorely disappointed with the outcome. If they did so, they would have grown out a segregating F2 population of

---

[‡]Where according to the university website *"he leads a research program using transcriptomic, proteomic and metabolomic approaches to understand the effects of Transcendental Meditation on gene expression"*.

genetically diverse maize with a wonderful trait that plant breeders call "yield resistance". Reserving seed for the next year's planting would have been something that the farmer would do only once, and this poor genetic outcome would have had nothing to do with the GM nature of the plants. It's called inbreeding depression and genetic segregation, and would occur if you grew any seed harvested from F1 hybrid plants (see Chapter 2).

7.2 FARMER SUICIDES

This is one of those myths that the famous Indian anti-GM activist, Vandana Shiva, started and keeps embellishing on, even when faced with the real data. For a person whose basic training was in physics, she is remarkably cavalier with numbers and the interpretation of correlations. Among her many pronouncements are that:

1. Since the introduction of GM cotton into India, over 400 000 Indian farmers have committed suicide, although sometimes the figures change (250 000 or 370 000 or 310 000; www.navdanyainternational.it) depending on who is quoting them. And the fact that glyphosate has killed over 20 000 people in Sri Lanka is another good one doing the rounds.
2. The rates of childhood autism have been caused by the increased use of glyphosate (Figure 7.2) (which to her mind is also because of GMOs). And according to a number of websites,[§] this is what Indian farmers drink to commit suicide, particularly the Monsanto brand, which is always more expensive than the generic formulations.
3. GM crops are bad because they are monocultures. And yes, there are advantages and disadvantages of monocultures, but they have been around for centuries and have nothing to do with GM technology. In fact, if the farmer follows the contract and grows refugia of susceptible crops like pigeon pea, sorghum or conventional cotton, then Bt cotton will never be a monoculture and will increase biodiversity.

[§]Example: www.seattleorganicrestaurants.com are so confused they claimed that Bt crops meant farmers didn't have to buy Monsanto's "costly herbicide" any more and then resistance broke down so farmers had to buy the "costly herbicide" again.

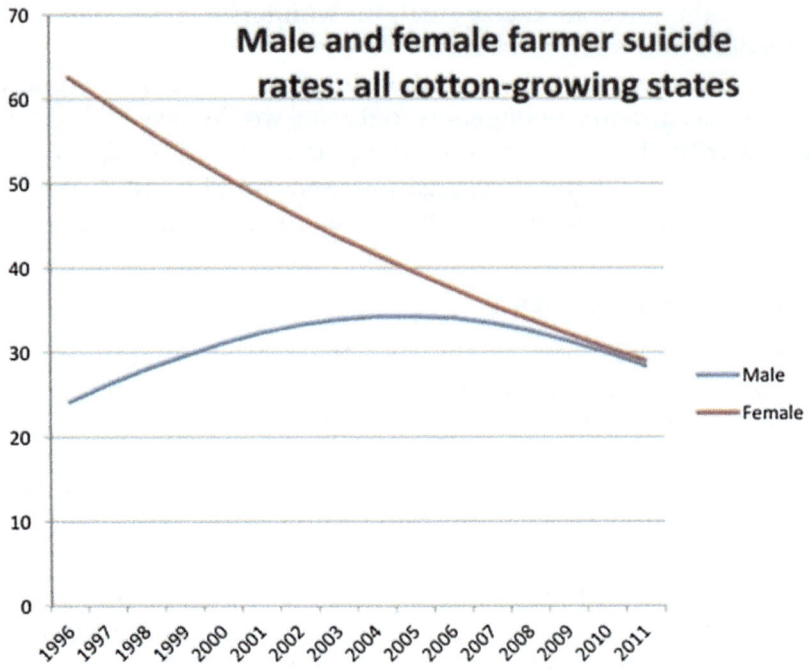

Figure 7.2 Male and female farmer suicide rates in cotton-growing states of India.
Image by Ian Plewis, Manchester (published in *The Conversation*).

4. Fertiliser is poison and should be banned. "It's a weapon of mass destruction. Its use is like war, because it came from war".[7] Presumably this comes from the fact that the Haber–Bosch process was used to make ammonia from nitrogen in the atmosphere. As discussed earlier, the demand for the process was driven by the need for nitrogenous fertilisers, and was well and truly working for that purpose by 1910, with the first industrial scale process by 1913. It is true that the production was diverted to the manufacture of explosives by Germany in World War One, when the British blockade made it difficult to source saltpetre in Chile for explosives. So while it is true to say that the process was used for warfare, that was not why it was developed, and agriculture still remains the main use of the ammonia produced by the process today.

The farmer suicides story has been propagated and spread by no less a luminary than Prince Charles, who stated in 2008 that "I blame GM crops for farmers' suicides" (*The Independent*).[8] Since he attended a conference in India in 2008 sponsored by Shiva, it has been reported that he keeps a bust of her at his house, Highgrove.[7] Then the claim was resurrected in 2012 in Michael Peled's film "Bitter Seeds".

For a scientific and numerate debunking of the myth and the numbers, one of the better sources is an article in *The Lancet* by Patel *et al.* (2012) from the London School of Hygiene and Tropical Medicine.[9] They showed that suicide rates have been static, and going down in recent years. As Ian Plewis from Manchester University showed (*The Conversation*), suicide rates in Indian farmers peaked in 2001, just before Bt cotton was introduced (Figures 7.3 and 7.4). Suicide rates appeared to be highest among farmers growing cash crops (which does include cotton), because farmers were most likely to be indebted if crops failed, usually because of drought or flood. Further analysis showed that suicide rates in rural areas were double those of urban areas, with the highest rate in Kerala (66 per 100 000, which was more than double the national average). Kerala has a long history of being

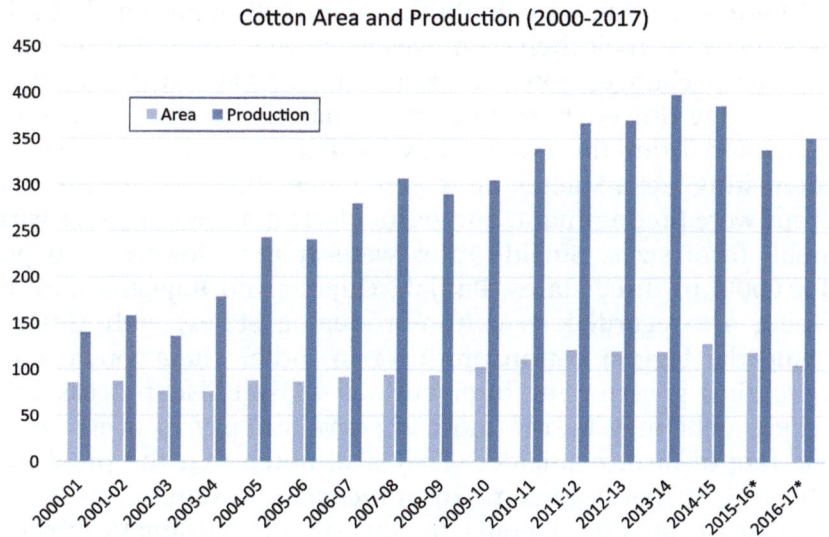

Figure 7.3 Cotton area and production in India from 2000 to 2017. Bt cotton was first made available to farmers in 2003.

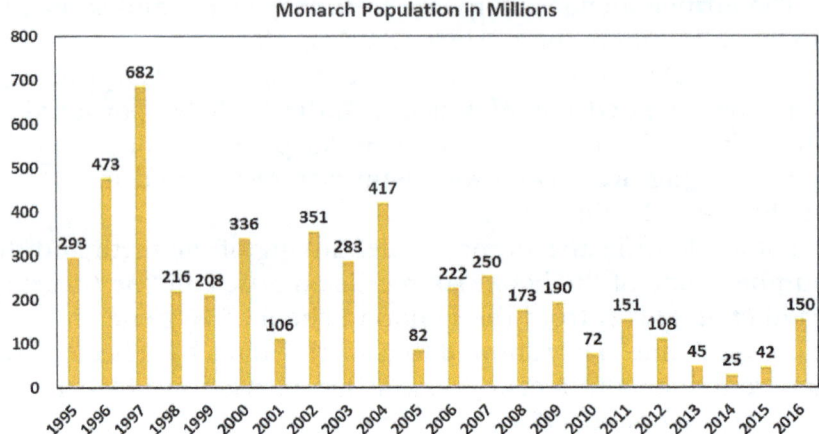

Figure 7.4   Monarch butterfly populations over 20 years (1995 to 2016). Figure by Tierra Curry, Center for Biological Diversity.

the epicentre of cash crops in India, particularly famous for spices (pepper, cardamom, cinnamon, cloves and nutmeg) and plantation crops such as coconut, tea, coffee, rubber and cashews. Kerala is very tropical with high monsoonal rainfall and is generally unsuited to the cultivation of cotton.

There is absolutely no doubt that some cotton farmers in India very tragically took their own lives when the crop failed or there were price changes. Jonathon Kennedy and Lawrence King from Cambridge University published an analysis in 2014, using data generated from the Patel study, which showed a number of interesting facts. Suicide rates were highest in states where cash crops were predominant, and particularly in marginal areas with small farm sizes. Suicide rates were at their lowest (<20 per 100 0000) in three states, Punjab, Gujarat and Rajasthan, all of which are regarded as cotton-producing states, with Gujarat being the biggest cotton producer in India. Their conclusions were that government interventions to stabilise process and relieve debt may be the most effective strategy to overcoming the rate of farmer suicides. They also noted that the pressures affecting farmers can come from increases in input costs (seed, fertiliser, pesticides, labour), but an even more dramatic effect is the volatility of the price farmers receive for their products. For example, in 2002, coffee growers were receiving $2.20 per kilo, and within four short years this had dropped to 40 cents per kilo. How would you fare if you suddenly took a pay cut of over 80%?

What we can say is that the data on farmer suicides showed that the rates in India were about the same as the rates in Scotland and France. In most of the cotton-growing states, suicide rates were lower among farmers than in non-farming males. There is also evidence to support the hypothesis that suicide rates in the cotton-farming states have been declining since 2005.[10]

7.3 BT MAIZE KILLED MONARCH BUTTERFLIES

Yes, Bt maize pollen has been demonstrated to kill Monarch butterfly larvae. Monarch butterflies (Figure 7.5) are a totemic species, and known the world over for their seasonal migrations, most famously in North America, because the Discovery Channel was there to film it. OK that's not the only reason. It is an impressive annual event in autumn when the Monarch populations of North America fly south for the winter, with the East Coast and Midwest populations heading south to Florida or Mexico, and the west coast ones flying south to overwinter in Mexico. The same species is all throughout the Pacific and I know from

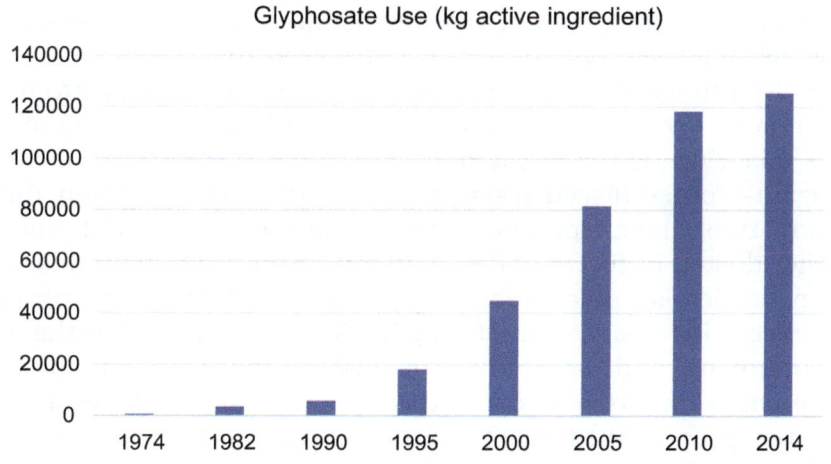

Figure 7.5 Data in thousands of kilograms or pounds of glyphosate active ingredient. From the National Agriculture Statistical Service pesticide use data and the Environmental Protection Agency pesticide industry and use reports (1995, 1997, 1999, 2001, 2007).

personal experience that the same species are in Australia, although I was always taught to call them Wanderers. They undergo migration flights here too, although on the East Coast that mainly consists of leaving the inland regions to head to the coast where the night temperatures are much milder in winter.

For many years, scientists have observed the annual migrations, in part to monitor the population. In Mexico, the estimates are taken of the actual total area inhabited by the overwintering colonies, and numbers extrapolated based on these areas. This is collated by groups such as WWF Mexico and released to the media towards the end of winter. Similarly, estimates of population sizes are made in summer in the Midwest, and hence an overall picture of the population size can be arrived at. It is a huge amount of work, and the groups doing the surveys do an amazing job to generate meaningful data. Yet the numbers can be difficult to really synthesise into what is happening with the population. In the 1990s there was concern and some evidence that there was a decrease in the population, and some research started to link this to GM crops, specifically to Bt maize.

A group of researchers at Cornell University, led by entomologist John Losey, undertook a study to test this. Monarch larvae are very fussy eaters, and do not see maize plants as a food source. In fact, they are so picky that the species will only eat milkweed, which includes mostly plants of the genus *Asclepias*, which has over 140 species. It is believed that Monarch larvae evolved to eat these plants because the milky sap they produce contains toxic compounds called cardenolides. Hence the toxin is taken up by the larvae and gives them a level of protection against other insectivorous animals. The Cornell group then undertook their experiments under lab conditions and found if they dusted Bt maize pollen onto milkweeds, then placed larvae on the leaves, there was a high level of mortality among the larvae. This was it – proof that GMOs were a hazard to the environment, and were going to lead to the extinction of the Monarch butterfly. Media and environmental groups went into hyperspace. By the time they returned to the Earth's atmosphere, the study had been scientifically discredited. In fact, the counter-study was so comprehensive that a group of six scientific papers was published as companion articles in the *Proceedings of the National Academy of Science USA* in 2001.

Nobody disagreed that pollen containing Bt toxin could kill Monarch larvae in sufficient dose. Remember, everything is toxic, it all depends on the dose. Was the heavy dusting of pollen likely to be experienced under real-world situations, and would female Monarchs lay eggs on these dusty yellow leaves, or would they choose to lay elsewhere? I recall a pundit at the time saying that if you totally covered the leaves of their preferred food, milkweed, with pollen "as if it was parmesan cheese", it was no surprise that there was enough pollen to be toxic. It was also not a surprise that many media articles linked this to Monsanto. As we've already seen, there have been instances where Monsanto have perhaps done the wrong thing. This is not one of them.

The most widely used Bt maize "event" at the time was a Monsanto product, MON810, and as we have already discussed the Cry9C StarLink maize of Aventis was also widely cultivated. The experiments involving Bt pollen and Monarch butterflies used a Novartis event 176. This was a good choice for this experiment, because Bt expression pollen was around 10-fold higher in event 176 than it was in other Bt maizes. It was also only grown on less than 2% of maize area planted in the USA.[11] Using event 176, Stanley-Horn *et al.* (2001) looked at the effect of dosage of pollen, in terms of number of pollen grains per cm^2 of leaf surface.[12] They found that at 22 grains per cm^2 there was no mortality of larvae, but there was a small but noticeable effect on larval growth rate. At the rate of 67 grains per cm^2, there was significant insect mortality (60% of larvae died).

How did these rates stack up against what was observed in the field? Milkweed tends to favour disturbed areas, hence it can be at high populations within a maize field and especially in the periphery of an agricultural field. Surveys in a number of fields showed that within a maize field, during the time of pollen shed (providing it had not rained), there was a mean of around 171 grains per cm^2.[11] This was definitely sufficient to cause insect mortality. They also found that because maize pollen is quite heavy, there was a 5-fold reduction in pollen densities 2–3 metres from the edge of the maize planting. Hence on average, pollen density outside the maize area was around 37 grains per cm^2, sufficient to cause slower larval growth rates.

However, the amount of Bt protein in MON810 and StarLink (Cry9C) pollen was 10-fold lower compared to event 176. Hence it was highly unlikely that the cultivation of maize containing the MON810 and StarLink events would have had any adverse effects on Monarch butterfly numbers. It was also true that the density of milkweed was approximately four-fold higher in non-agricultural lands – pastures, roadsides, cleared areas. However, that in itself may be misleading. Monarch butterflies tended to favour the milkweed plants in maize fields, as agricultural fields seemed to be responsible for about half the Monarch numbers.

So I think we can agree that based on these figures, Monarch butterfly numbers were not adversely affected by Bt maize. A look at the actual numbers also suggests that during the time of these studies, Monarch numbers actually went up. To some, this was a justification that Bt maize was better for the survival of Monarch butterflies. The argument went that Bt maize was much more environmentally benign than the spraying of pepmethrin, bifenthrin, lambda-cyhalothrin and methylparathion to control European corn borer. It is worth noting that losses of $1.2 billion of maize production per annum was also a terrible waste of resources. Nothing to see here. More fake news. I should also point out that in the study that precipitated all this furore, Losey and co-workers stated:

"It would be inappropriate to draw any conclusion about the risk to Monarch populations in the field based solely on these initial results."

Yes, fake news. And I myself am responsible for fake news right here in this chapter. In a previous paragraph, I stated that Losey's study was discredited. This is not the case. The study was well conceived and performed. It demonstrated that in maize lines which had significant Bt toxin concentrations in the pollen, ingestion by Monarch larvae could lead to death. It was subsequently demonstrated that, at least in the case of event 176, pollen density required for mortality can be physically attained in the field, although this dissipated rapidly a few metres away from the maize. Therefore, you could draw the conclusion that if event 176 was the only source of Bt maize in the USA, there may have been quite serious implications for Monarch butterfly numbers.

Figure 7.6 *Danaus plexippus* (Monarch butterfly).
© Kate Besler/Shutterstock.

A look at Monarch numbers over the time period shows that they are quite variable (Figure 7.6). From the two years from the time of the first study (1998) to the completion of the second study (2000), many people pointed out that butterfly numbers were going up (from 216 to 366 million). While that is true, it did not account for the huge collapse of numbers from 1997, when numbers were estimated at 682 million. Overall there was a long-term average of around 300 million, and numbers went up or down depending on climatic conditions, particularly drought or freezing conditions.

However, if you look at the past decade (2008 onwards), numbers have never climbed above 200 million, and have been below 100 million for four of those years. Most people would look at those figures with an element of concern. Yes, climate change has led to greater uncertainty, and without a doubt that is a contributing factor. However, the Centre for Biological Diversity has put forward the hypothesis that the increasing use of herbicides could be a major driver for population numbers going down and staying down, due to reductions in their food source. The issue does get somewhat confused however by statistics. The numbers of Monarch butterflies are based on the overwintering populations in Mexico. These are based on the main sanctuaries.

How does a person go and count Monarch butterfly numbers? The short answer to that is, in fact, they don't. What is done is a measure of the size (in terms of area) of the overwintering population is made, and then extrapolated out based on numbers of butterflies per acre. To generate useful data, it is not generally accurate to base this on measurements in one site, because each site can be quite variable, for reasons not well understood.

Table 7.1 shows that there was about a 27% decrease in numbers overall in Mexico. Had the sampling only been performed in El Rosario, the conclusion would have been that numbers were steady or slightly increasing. If only performed in Cerro Pelon, the conclusion would have been that there was a catastrophic drop in population (data from JourneyNorth.org). Hence, while Monarch numbers were deemed to be remaining constant in surveys in the Midwest, they appeared to be declining in surveys in Mexico. Hence, either there was an increase in mortality among butterflies migrating from the Midwest and never reaching Mexico, or the numbers in one (or both?) of the places was incorrect.

A recent study involving an international team including one of my colleagues, Professor Myron Zalucki, an entomologist and Fellow of the Australian Academy of Sciences, has raised this issue.[13] This study shows some compelling evidence that from 1999 to 2009, milkweed populations in some of the Midwest agricultural states such as Iowa and Illinois have declined by 97% and 94%, respectively. It is also the case that in these states, 92% of the agricultural land is soybean and maize and that most of that cropping area is planted with herbicide-resistant cultivars of these crops. The main outcome of this is that Monarchs have been forced to move from agricultural to non-agricultural lands. The surveys undertaken to determine that numbers were not dwindling were undertaken in non-agricultural land areas, and these counts suggested that numbers were not going down and

Table 7.1   Monarch butterfly numbers recorded at two sanctuaries in Mexico demonstrating local variation from year to year.

Sanctuary	2015–16	2016–17
El Rosario	55 million	59 million
Cerro Pelon	47 million	16 million
TOTAL Mexico	200 million	146 million

may even have been increasing. There is good evidence to suggest that the Monarch populations have effectively shifted from agricultural land (where around 50% of the total population once lived) to the non-agricultural land. The move has been driven by the effectiveness of herbicides at limiting weed numbers, which, of course, includes their favourite source of food, the milkweed. Pleasants and co-workers reached the conclusion that the milkweed limitation hypothesis for Monarch decline is supported by their analyses, and that efforts to increase milkweed populations in the landscape across the summer breeding region have a sound scientific basis.

Hence, while it is fairly clear that Bt corn did not have any adverse effects on Monarch populations, the increasing use of herbicides did. In highly agricultural areas where herbicide-tolerant crops such as maize and soybean make up an overwhelming proportion of agricultural land, it can clearly be argued that herbicide-tolerant crops have played a role in the decline of Monarch butterfly numbers. Hence the anti-GM activists were barking up the wrong tree (or to be botanically correct, the wrong corn stalk in this case).

7.4 GMO = PESTICIDES

Yes, a GMO can be a pesticide. Bt crops, for example, can require registration by the same bodies that register pesticides. This is certainly the case in Australia, where the Australian Pesticides and Veterinary Medicines Authority is required to give clearance for crops which produce Bt toxin as it is classified as a pesticide. The outcome of that is you have crops such as Bt cotton which significantly reduce the need for pesticides.

And yes, you could also use that descriptor to describe Roundup Ready or other herbicide-resistant crop systems, because the crop is designed to have that herbicide applied to reduce the pressure of weeds.

However, Flavr Savr tomato is not a pesticide. The recently released Arctic apples, which don't go brown when you cut them into slices, are not a pesticide. The potatoes which have early blight resistance are not a pesticide (and as we shall discuss, will lead to hugely reduced fungicide applications). The sorghums my lab has produced, which have larger grain size and higher protein contents, are not a pesticide. The faster growing Aqua-Bounty salmon are not a pesticide. And like most applications of GM, these and most other GM plants and animals are not pesticides and do not require the use of pesticides to deliver the yield and quality advances they are designed for. Hence, GMO $\neq$ pesticides.

7.5   GMO IS NOT ORGANIC

We will discuss this one at length in the next chapter. However, for the sake of things here, I draw your attention to the definition of organic in the Oxford English Dictionary (the Abridged, the Concise, the one you used at school in 1953 – it doesn't matter).

"Relating to, or derived from living matter."

Hence, what is organic food? Can you think of a food that is not "derived from living matter"? Hence I think the family restaurant that sells foods that contain "two all-beef patties, special sauce, lettuce, cheese, pickles and onions on a sesame seed bun" would be using the word correctly if they were to refer to this food as organic. Surely pretty much all those ingredients are "derived from living matter?" Not convinced about the "special sauce"? Well I have to say I'm not convinced about the cheese, which does seem to do a particularly good imitation of yellow plastic. Apparently there are a lot of websites which claim to have either found the recipe for the special sauce or decoded it, and given that they all have slight variations to the recipe, I will leave this research up to you. It's right up there with the "11 secret herbs and spices" of KFC – a good marketing gimmick. Even the Coca-Cola recipe is super-secret, but parts of it are organic. Sugar. Sugar is derived from plants, most commonly either cane or beet. All food is organic.

However, not all food is "certified organic". GMOs are not "certified organic". That's because they are "not natural". Apparently spraying large amounts of copper sulphate or petroleum oil or rotenone on a crop is natural. Also totally natural is taking a soil-borne bacterium, producing a purified culture of one type called *Bacillus thuringiensis*, growing it up in a factory in large batch cultures so that it multiplies billion-fold, centrifuging or filtering the cells, freeze drying the cells and putting them into hermetically sealed packages with powdered talc (or in some cases 43% trade secret ingredients?), shipping them all over the world, then reconstituting them in water and spraying them on a crop, or dusting them onto the crop. I mean, really, nature just doesn't get any better than that does it?

So, I will curtail this discussion until Chapter 8 and leave you with this thought. **GMOs are organic.** Just like all lettuces, tomatoes, duck breasts, smoked elk, vegan kale and chia milk (is that a thing?), deep-fried pizzas and beer. Beer is still organic even if you add burnt orange peel, coriander, rosehips and mango. The fact that GMOs are not "certified organic" is an entirely human construct. A human construct that could one day be easily reversed, if somebody, somewhere, decided sometime that this would be a good idea for agricultural sustainability. I think it would be a good idea, and I'm not alone.

REFERENCES

1. J. Randall, Prince Charles warns GM crops risk causing the biggest-ever environmental disaster, The Daily Telegraph; Tuesday 12th 2008, https://www.telegraph.co.uk/news/earth/earthnews/3349308/Prince-Charles-warns-GM-crops-risk-causing-the-biggest-ever-environmental-disaster.html.
2. C. J. S. Smith, C. F. Watson, J. Ray, C. R. Bird, P. C. Morris, W. Schuch and D. Grierson, Antisense RNA inhibition of polygalacturonase gene expression in transgenic tomatoes, *Nature*, 1988, **334**, 724–726.
3. A. Hamilton, G. W. Lycett and D. Grierson, Antisense gene thatinhibits synthesis of the hormone ethylene in transgenic plants, *Nature*, 1990, **346**, 284–287.
4. W. Schnippenkoetter, C. Lo, G. Liu, K. Dibley, W. L. Chan, J. White, R. Milne, A. Zwart, E. Kwong, B. Keller, I. Godwin, S. G. Krattinger and E. Lagudah, The wheat Lr34 multipathogen resistance gene confers resistance to anthracnose and rust in sorghum, *Plant Biotechnol. J.*, 2017, **15**, 1387–1396.
5. M. Kuntz, The postmodern assault on science: If all truths are equal, who cares what science has to say?, *EMBO Rep.*, 2012, **13**(10), 885–889.

6. J. Green, *Moral Tribes*, Penguin Press, 2013.
7. M. Specter, Seeds of Doubt, New Yorker Magazine, August 25, 2014.
8. G. Lean, *The Independent*, 4 October, 2008.
9. V. Patel, C. Ramasundarahettige and L. Vijayakumar, *et al.*, Suicide mortality in India: a nationally representative survey, *Lancet*, 2012, **379**(9834), 2343–2351.
10. I. Plewis, Indian farmer suicides: Is GM cotton to blame?, *Significance*, 2014, **11**, 14–18.
11. M. K. Sears, R. L. Hellmich, D. E. Stanley-Horn, K. S. Oberhauser, J. M. Pleasants, H. R. Mattila, B. D. Siegfried and G. P. Dively, Impact of Bt corn pollen on monarch butterfly populations: A risk assessment, *Proc. Natl. Acad. Sci.*, 2001, **98**(21), 11937–11942.
12. D. E. Stanley-Horn, G. P. Dively, R. L. Hellmich, H. R. Mattila, M. K. Sears, R. Rose, L. C. H. Jesse, J. E. Losey, J. J. Obrycki and L. Lewis, Assessing the impact of Cry1Ab-expressing corn pollen on monarch butterfly larvae in field studies, *Proc. Natl. Acad. Sci.*, 2001, **98**(21), 11931–11936.
13. J. Pleasants, S. R. Leather and A. Stewart, Milkweed restoration in the Midwest for monarch butterfly recovery: estimates of milkweeds lost, milkweeds remaining and milkweeds that must be added to increase the monarch population, *Insect Conserv. Divers*, 2017, **10**, 42–53.

CHAPTER 8

Not Ready to Make Nice

"The greatest derangement of the mind is to believe in something because one wishes it to be so."

Louis Pasteur

"What is food to one man is bitter poison to others."

Lucretius

Celery. That's right – I said celery. If you've been paying attention, you may be forgiven for formulating an opinion that I have something against celery. You would be right, and I have the scars to prove it. No that's not true anymore, but I did have the scars over 30 years ago, which faded after a couple of years. Nevertheless, I do have a strong feeling that celery is an example of one of nature's "bitter poisons".

While doing my undergraduate degree in Agricultural Science, one summer I spent four weeks on a celery farm on Queensland's Granite Belt near Stanthorpe. Much of the farming around Stanthorpe is in a former "soldier settlement" area where the wisdom and magnanimity of the Australian government aimed to help soldiers returning from World War One to gainful employment by granting them a 100-acre farm. These soldier settlements were usually not "farms" as such, but mostly areas of virgin land, commonly forest. We Australians would call it bush (if the trees were tall) or scrub (if the trees were more shrub-like). These farm

Good Enough to Eat? Next Generation GM Crops
By Ian D. Godwin
© Ian D. Godwin 2019
Published by the Royal Society of Chemistry, www.rsc.org

allotments were granted to returned servicemen on the basis that they had to clear it (cut down all those useless trees) and turn it into productive farmland. If you've ever read A. B. Facey's Australian classic *A Fortunate Life* you will understand the predicament these men were placed in. I cannot recommend this book highly enough. It has nothing to do with GM crops or organic crops, and it is definitely celery-free, but it illustrates the rigours of life on the farm, showcasing the tough choices farmers need to make in the face of social, economic and environmental realities.

Back to the celery farm. It was (and still is) in a locality that became known as Amiens, which some of you will know is a town in northern France. It's known for being the birthplace of Jules Verne and Emmanuel Macron. Many will also know it was of strategic importance to both sides in 1914–18, not least of which because it straddles the river Somme. In nearby Amiens (the Queensland one), you can drive along Messines Road or Bapaume Road to get to Passchendaele State Forest and Pozieres. There is no Somme, but there is Cannon Creek, so I think you get the picture. The family who ran the celery farm were a wonderful, generous family, the Harsletts. Every summer they welcomed significant numbers of students and international backpackers to do the harvesting. They provided nice huts to live in, and a veggie patch with a pick-your-own pretty much everything. In reality, it wasn't just a celery farm, as they also grew lettuces and broccoli. Every morning the rag-tag group of workers would rise before dawn and fuel up with energy-rich food and coffee with the beautiful fresh milk from their own cows. I don't want to sound ungrateful, but there was no two ways about it, the milk was fresh, but it wasn't beautiful. This natural unpasteurised milk was the worst thing you could ever put on your corn flakes. Not because it was unpasteurised, but the free-range cows grazed freely on paddocks that contained a weed (I think it was *Coronopus didymus*) that imparted a pungent sulphurous odour and made the milk taste like somebody had put a spoonful of mustard into the container. You could still smell it hours later on your breath. Then after breakfast you would go out into the cool of dawn, muster around the packing sheds, and wait to be assigned your day's work. If you got broccoli – awesome. You got to stand up beside a conveyor belt, and trim the inflorescences as they went by, on their way to the polystyrene packing boxes. If you got lettuce, not so good. Bending down to cut the lettuce at ground level and twisting them pairwise into boxes you had to stack on a truck. If your name wasn't called out, you were with the rest of the crew on celery all day. This was far worse than lettuce.

Celery cutting was hard work. It involved a really sharp knife that had to be applied expertly and with conviction to cut in exactly the right spot (including not your own fingers). Cut too high and the stalks separated (this was not good and the boss always noticed, and let you know they noticed). Too low and you cut into the granitic "soil", and had to continually resharpen your knife (see note about boss above). Then there was lots of heavy carrying. Yes, cutting celery was most definitely negative calories. Then there was the inescapable pain. Celery juice is without a doubt toxic in the true sense of the word, or at least that's what we all thought. After the first day of cutting celery, you would have blisters start to develop all over your forearms, for most people with a burning sensation. Your hands were gloved up, but that was the extent of the personal protective equipment in 1982. When you carried your completed tray of celery to the truck, there was no way to do it without effectively hugging the 15 kg of celery to your chest, and the terrible juice leaked all over you. Then the juice started to react with the ultraviolet radiation (sunlight) and blisters started to form. Then for the next four weeks the blisters would be bursting and re-forming all over your arms, while your skin became more and more unsightly and painful. You often got some very lovely patterns, caused by the indentations made by the parallel vascular bundles in those extended petioles (or what we call celery stalks). On my lilywhite Anglo-Saxon/Celtic skin, those marks were still visible two years later. A purple and red pastiche of lines and blobs, which in hindsight was very Jackson Pollock and could have been a social media sensation.

Celery juice contains psoralens and other furanocoumarins (just in case this is not clear, these are "natural" and "organic" chemicals). Furanocoumarins have the effect of increasing your skin's sensitivity to UV light. There was no meaningful way of avoiding the sap. In Southern Queensland where the summer sun is UV intensity 11 + (extreme) from about 9 am onwards, it was difficult to come up with a satisfactory means to avoid the sun. What the psoralens actually do is react with DNA, in a reaction that is activated by UV photons, the outcome of which can be mutations to your DNA. Celery dermatitis was first reported in scientific literature in 1926 in France,[1] and became a widespread phenomenon in the USA ten years later when a new variety, Pascal, was introduced. By 1939, Pascal was the predominant variety, and farm workers reported major problems with adverse skin reactions. A combined study by the Massachusetts General Hospital and the Agricultural Experiment Station of the University of Massachusetts reported variation. The variations were in the reactions by the workers, and also dependent on the age of development of the celery.[2] Interestingly they noted "dark-skinned laborers from Jamaica and the Bahama Islands employed on the farms seemed to be immune". They concluded with the statement that "clinical evidence indicates a special dermatogenic quality peculiar to the green summer Pascal celery". In fact, it was the arrival of over a million US servicemen in Brisbane in the early 1940s that turned the Harslett's into a celery farm. Prior to that, there was not much demand for the vegetable. Sadly, in the past few years the family has had to lease the farm to larger corporations and they no longer farm celery.

In the 1980s, it was documented that a new insect-resistant celery variety released in the USA contained almost eight times the furanocoumarin content as previous varieties. This variety, like all other celery varieties, was produced by conventional plant breeding. Reports of adverse skin reactions were investigated in grocery store workers[3] and it was also shown that among farm workers, 100% of those surveyed reported not only the skin effects, but many suffered from "residual hyperpigmentation", meaning the marks didn't go away. We also know that celery is not alone in this wonderful feature, because the same effects can be induced by figs, carrots and parsnips. There is even an effect known as margarita photodermatitis caused by drinking margaritas outside (and who in their right mind would consider drinking margaritas inside). If you squeeze too much lime juice onto your hands and not enough in the actual glass, wash your hands immediately. What is interesting (at least to a geneticist) is that there is biological variation – both within the plant species and the human species. Some celery does not seem to induce dermatitis, presumably because it has lower levels of the offending biochemicals. Some humans seem immune to the dermatitis, even when the levels of furanocoumarins are almost 10-fold higher. Hence like all things in nature, dosage is important when it comes to toxicity. It is also true that environment is a major factor, and in this specific case there are no apparent ill-effects of the furanocoumarins in the absence of sunlight.

However, as we have already discussed, one woman's toxin is another's tonic. Psoralens are used to treat various skin disorders and may also have some efficacy against certain types of cancer. That's biology (and chemistry). Everything good can be bad. Everything bad can be good. This is nature. "Natural" by its very definition is not safe. Just like animals. We know that and we have an innate sense of this natural danger. We swim in the ocean and have a background fear of sharks and marine stingers. We walk through forests, depending on where we are, and we keep a wary eye out for bears, snakes and tigers. Animals kill people. Most effectively, mosquitoes are the ultimate killing machine, almost entirely through the viruses and microbes they carry, with malaria alone estimated to kill 600 000–750 000 humans per annum. That's 600 000–750 000 more than GM plants.

Plants also kill people. And they are entirely organic. Some are even entirely "certified organic". In saying that, I don't think celery has ever killed anyone, despite the fact that there are various nutritional websites saying how toxic celery is. Probably about the same number as there are websites listing how healthy celery is. I don't know what "science" truly says, but I can say that my "values" dictate that celery should be viewed with suspicion, and taken in moderation. Twice a year at most, preferably disguised in a casserole, or if absolutely necessary, smothered in a taste-concealing dip. It doesn't take away the danger you know. There have even been documented cases wherein people have eaten celery soup, then gone out into the sun or to the tanning salon and very rapidly succumbed to sunburn.[4]

When it comes to killer plants, one in particular is the biggest killer of all – tobacco. We all know it produces large amounts of toxins, and we also know that chewing or setting the dried leaves alight and inhaling the smoke is a great way to shorten your life. Yet we still sell tobacco leaves in many forms in stores on every corner. According to Cancer Research UK, there are over 14 million new cancer cases every year worldwide (Cancer Research UK) and approximately one-third of all cases can be attributed to exposure to tobacco smoke. For some, their "values" dictate that they can find ways to ignore these risks. Very commonly their "values" are ably assisted by advertising featuring rugged men on horses, or beautiful, smart women whose sophisticated lifestyle is made possible by smoking a particular brand.

> *"When I'm watching my TV, and a man comes on and tells me,*
> *how white my shirts can be.*
> *But he can't be a man cos he doesn't smoke,*
> *the same cigarettes as me."*
>   The Rolling Stones, Satisfaction (I Can't Get No), 1965

For others, their "values" lead them to conclude that tobacco should be banned in all its forms. Then there is everyone else in between. That's the nature of human values. In recent years we will very probably be diversifying the source of these deaths by adding marijuana to the list of legally sold toxic substances.

When it comes to human values, many of us on the planet strongly believe that we should be doing many of our activities differently. Among those activities is the business of food production. We all eat food every day. And, despite for many of us our urban existences, we are all aware that a vast majority of the Earth's surface is taken up with the production of our food. Hence when it comes to a footprint on the planet, food production is a pretty big one. While writing these words, I am living and working in Copenhagen. When I go to do the groceries with my lovely wife, nearly every item we select decries that it is *Økologikal* or *Øko* or *Oeko*. The dictionary definition of ecological is "relating to or concerned with the relation of living organisms to one another and to their physical surroundings". A similar term in usage in the supermarket in Germany was *Öko*, meaning much the same thing and one could, at least somewhat within a scientific framework, have some understanding that this food was produced using ecological principles. However on other foods the description was *Bio*. Presumably *Bio* means biological, and like the term "organic" has taken a term with a meaning and restricted it to mean something quite different. Your food is all biological and it is all organic. What these labels mean is that food is produced under a proscriptive regime where a select subset of chemicals may be used. These chemicals are, naturally (well, there I go misusing that term), not called chemicals, but are known as "allowed inputs", or many other things. They are, however, chemicals. However, I shall come back to the "Allowed Inputs" (Australian Certified Organic), the "National List of Allowed and Prohibited Substances" (USDA Organic), "Standards" and "Exceptional Permissions" (Soil Association Organic), or "Permitted and Restricted Fertilisers" and "Allowable Materials and Methods for Plant Care" (Demeter Bio-Dynamic), in a few pages' time. These lists include the chemicals that "certified organic" or "biodynamic" producers are allowed to use.

Surveys from the USA (Pew Research Institute, 2016) showed that 76% of people who bought organic foods did so because they believed they were healthier. In the same survey, 39% of Americans believed that GM foods are worse for their health. Looking across various surveys, the most common responses from people who believe organic food is healthier believe this to be the case for one or more of the following reasons:

1. It is natural or "as mother nature intended".
2. It is chemical-free.
3. It is grown by small farmers who live locally, so the food is fresher.
4. It is better for the environment.

Some people said they choose organic food specifically because they want to avoid GMOs. Without exception, one of the things that every organic certifier has in common is that they preclude the production of GM plants, or the use of GM plant-derived products as animal feed. This is a situation that appears to be accepted as "cast in stone". But it is also true that back in the 1990s, this was not necessarily a certainty.

The current USDA Organic lists GM products as prohibited substances, so you can't sell an apple as organic if it is GM. You can't feed GM maize to a cow that is to be marketed as organic beef or producing organic milk and cheese. You can't use GM soy lecithin as a food additive in your organic chocolate. I was in the USA in the 1990s, visiting Peggy Lemaux and Bob Buchanan at University of California, Berkeley, when the debate was raging in the USA. When I say raging, it was on National Public Radio.

Most people didn't care less, and at that time, the certified organic movement was not something that had many people's attention. USDA Organic was to be the established labelled certifier, and they were debating whether GM crops, which at the time only included maize, cotton and soybean, could be regarded as organic.

There were also two very separate classes of GMO. One was engineered to be resistant to the herbicide glyphosate. No sane farmer would pay the extra to buy herbicide-resistant seed, and then not use the herbicide to control weeds. Glyphosate was not a herbicide that organic growers were allowed to use, hence there was no prospect that GM Roundup Ready maize or soybean was ever going to be grown to certified organic standards. Glyphosate was man-made and therefore not allowable under the "allowable inputs". To control weeds you had to use manual labour, plough the field numerous times, or spray them with natural things like salt or acetic acid, which as you can understand are much less toxic and more environmentally sustainable than glyphosate (note this is sarcasm for those readers who are unable to detect it). Perhaps your "values" tell you that salt has to be less toxic than glyphosate. However, the science of toxicology will tell you otherwise.

One of the most widely recognised tools of toxicology is the lethal dose 50 or median LD_{50}. This involves calculating what dosage of a particular substance (which could be a chemical, radiation or even a pathogen) that is required to kill half the population. I can just imagine you reading this and saying "yes, that's it, I want to be a toxicologist". Like all sciences, it's not particularly straightforward. The LD_{50} is calculated on a body mass weight, and hence is expressed as how much of the substance (for example in milligrams) is required for toxicity per unit body mass (usually in kilograms). And then you need to know how the dosage is administered, whether it be oral, through skin contact or inhaling the vapour or gaseous form. If you take a substance like salt (NaCl), it has an oral LD_{50} of 3000 mg kg^{-1} for rats. That means if the average rat weighs 1 kg, 3 g (3000 mg) of salt administered orally is sufficient to kill half the population. If you extrapolate that for humans, and take a small human, like a 50 kg jockey, that means you need 150 g of salt for half the population to be killed.

So why half the population? Well a major reason for that is genetic diversity. Some individuals may be more susceptible and some considerably less. It's not dissimilar for pharmaceuticals. Many of us know individuals who say taking two paracetamol tablets has no effect on their headache, whereas others can take half a tablet and fall asleep for 12 hours unable to move. Yet genetic diversity is also problematic when it comes to extrapolating between species.

Recall that salt has an oral LD_{50} for rats of 3000 $mg\,kg^{-1}$, so we can extrapolate that it would be similar across mammals, of course including humans. Well, not just no, but emphatically NO. For mice, which are not too genetically distant from rats, the oral LD_{50} is 4000 $mg\,kg^{-1}$. So, on average, mice are more salt tolerant than rats. And all pet owners have probably heard that chocolate is dangerous for their cats and dogs. The toxic component of chocolate is theobromine. The oral LD_{50} for cats (200 $mg\,kg^{-1}$) and dogs (300 $mg\,kg^{-1}$) is considerably lower than humans (1000 $mg\,kg^{-1}$), hence while a few squares of chocolate have some great beneficial effects for you, they could be life endangering for your dog. Of course, there are always exceptions. I once posted (as in *via* the mail, not online) my sister some expensive liqueur-filled chocolate from Switzerland. The parcel duly arrived at her house in Brisbane. Her Jack Russell terrier, the much-missed Pep, found it before they could open it, ripped open the package and devoured two entire chocolate blocks along with some of the wrapping. They raced Pep off to the vet but he was not affected at all. The next day he had a hangover and a lifelong craving for cherry brandy. Perhaps this confirms one of my long-held suspicions that Pep was actually a rat (oral LD_{50} 1265). Even worse, my sister's current vet (I think she doesn't actually like dogs), insists that dogs don't need meat and that celery is a very useful and healthy dog treat. Celery? Yes, I know. I have this thing about celery, but to my mind this constitutes animal cruelty without a doubt.

Below is a list of some of the "allowable inputs" for Australian certified organic farmers, and their LD_{50} in rats. Farmers are allowed to use these on their certified organic crops (as sprays or soil applications) with few restrictions. Many are also substances we regularly and willingly ingest every day.

Substance	Oral LD_{50} in rats
Rotenone (derris dust)	60
Caffeine	127 (mouse)
Copper sulphate (as fungicide)	300
Lime sulphur (as fungicide)	820
Zinc sulphate (as fertiliser)	1260
Theobromine (chocolate)	1265
Magnesium sulphate (Epsom salts)	2000
Salt (NaCl)	3000
Acetic acid (vinegar)	3310
Glyphosate (active ingredient Roundup) NB not an allowable input	>5000
Homeopathic remedies (no active ingredient)	No active ingredient Same as water (which is >90 ml kg^{-1})[†]

Now you will see here that homeopathic remedies, as well as being the most expensive remedies per unit of active ingredient (because dividing by zero is infinity) can be, and are, more toxic than many other substances. That is because they contain water. Water can kill you, and not only by drowning (which is asphyxiation, not toxicity) or physical force (as experienced by huge wave surfers). Let's make one thing really clear, the homeopathic remedy may have deleterious effects on your disposable cash, but it is not going to harm you, unless you're waiting for it to cure cancer. The oral LD_{50} for water is 90 ml kg^{-1} of body weight. Hence that 50 kg small adult we talked about earlier would need to ingest >4.5 litres of water very quickly to get to the LD_{50}, which is quite difficult to do under most circumstances. Just as it is quite difficult to swallow enough salt to approach the LD_{50}, which for that 50 kg human would be 150 g, without throwing up. While vomiting is unpleasant, this would have the beneficial effect of removing much of the salt ingested. However, there are quite a few documented cases of humans who have ingested too much water and died. It is called hyponatremia, because the outcome of drinking too much water

---

[†]Yes water can kill you, and not just by drowning.

is dilution of sodium (natrium) to such a low concentration that many cellular activities fail. Again, here is a great example of the biological nature of toxicity. Yes, too much sodium is toxic and can cause death. Too little sodium and death is also the outcome.

It's not the same for caffeine or theobromine, regardless of how often you may feel you would die without coffee or chocolate. There is no physiological requirement for either of these, however for many of us they are "feel good" chemicals. For others they are something to avoid, sometimes because of taste. Or you may be like my daughter, who gets heart palpitations from a single cup of coffee, unless it's decaffeinated.

It's also not the same for rotenone and glyphosate. We have no requirement for these, and would gain no benefit from them. However, one of these is 100 times more toxic and has been demonstrated to be carcinogenic. It's rotenone, and is also an allowed input in many certified organic agricultural systems, including in Australia. Rotenone is allowable because it is a "natural product", being derived from the roots of a number of plants, most notably from *Derris elliptica*, a leguminous plant, and jicama, the Mexican yam bean. The ground roots of *Derris* have been used as a powder for hundreds of years as a means to poison fish in many parts of Melanesia (including Papua New Guinea and Fiji). The powder is used widely in organic agriculture to control insects. As a home gardener and chicken owner, I have used it regularly, particularly to control ecto-parasites on my chickens. Then I did some research, and stopped using it. Not only is it fairly toxic to rats (LD_{50} 60 mg kg^{-1}), it is even more toxic to mice (LD_{50} 2.8 mg kg^{-1}). No mouse has ever weighed a kilogram, so that actually means for an average adult house mouse, weighing about 19 g, a mere 53 ng of rotenone would be lethal. Picture an empty, clean teaspoon. Beside it is a teaspoon with 53 ng of rotenone on it. They will be indistinguishable to the naked eye, or even the eye with reading glasses or a magnifying glass. And there I was, like many other home chicken enthusiasts, shaking it around the chicken pen with abandon under the knowledge that it was OK for certified organic so it must be safe. Hmm.

However, it is very pleasing to see that, like The Bible, some organic certification documents allow for the occasional revision. Rotenone (or derris dust), which is used as an insecticide in organic agriculture in Australia, has been banned in the EU and the USA since 2011. That's not just in organic agriculture, but in all agriculture. Switzerland reacted a little more slowly, finally banning rotenone at the beginning of 2014. This is not the case in places like Australia and Brazil. In 2009, when the motion to delete rotenone as a pesticide from the Codex Alimentarius Commission was proposed in 2009, the International Federation of Organic Agriculture Movements (IFOAM) opposed the motion. Therefore, just because you are eating organic food in the USA doesn't mean it is rotenone-free. Rotenone has stayed on the list of allowable inputs, and no imported certified organic foods are tested for its presence.

In my own, admittedly rambling way, I am illustrating here that glyphosate is a very safe and very low toxicity herbicide. Does it have effects on the natural ecosystem? Of course it does! It kills weeds. Weeds which are, by definition, just plants that happen to be in a place where we don't want them to be. Any human activity has an effect on the agricultural ecosystem.

There have been many attempts in recent years to demonstrate the dangers of glyphosate to humans. Unfortunately, some people's values get in the way of their ability to interpret scientific information that has shown there are no credible links between glyphosate and cancer in any form. However, in 2016, the International Agency for Research on Cancer (part of the UN World Health Organization), upgraded glyphosate from a 2B Possible Carcinogen to a 2A Probably Carcinogen. Just for your information, Group 1 the Known Carcinogens includes tobacco, alcohol, asbestos, diesel exhaust, ionizing radiation, processed meat, mineral oils (allowable as certified organic inputs), the rubber manufacturing industry and solar radiation (allowable and extremely advisable for all agriculture). As a Queenslander with pale skin, I have already had numerous carcinomas surgically excised from my skin, perhaps because I have on regular occasions sat in the sun, drinking beer and eating sausages while passively smoking, watching the diesel traffic and bouncing a tennis ball against the asbestos wall. Thank goodness I wasn't wearing a condom.

It's tempting here to just accept Dr Joe Jackson's Group 1 Classification and get on with having a life.

"Everything.
Everything gives you cancer.
Everything.
Everything gives you cancer.
There's no cure, there's no answer."

Joe Jackson, 1982

Raising glyphosate from a possible to a probable carcinogen created a political and scientific preparation 500 (that's shit in a cow horn – see Figure 8.1 below) fight. None of the other organisations looking at glyphosate found any evidence to support this and, predictably, anti-GM activists and chemical companies lined up in the trenches, with farmers and consumers caught in the middle. And for all those people who eat food, this was concerning. It was most concerning for those of us who love a good Aussie barbecue, with the required 3 kg of red meat per person, because red meat suffered the same fate as glyphosate. It was also concerning for the vegetarians. Their beloved tofu burgers were almost certainly from a GM soybean, and hence riddled with glyphosate. Unless it was organic, then it was probably riddled with the much more toxic carcinogenic chemical, rotenone, but nobody ever tested for that anyway.

Now, when I say the soybean was riddled with glyphosate, I'm guilty of the serious crime of hyperbole. The anti-GM activists, whose major values included hating Big Ag, especially Monsanto, and any form of chemical, especially one that is made by

Figure 8.1 Preparation 500, cow dung-filled horn to be buried in the soil during the cooler months.
© Shutterstock.

Monsanto, had made much of studies showing that glyphosate was contaminating everything. This contamination was especially of concern in soybean, given that most of the crop was now GM and Roundup Ready. A detailed study of glyphosate residues was published from work in Argentina,[5] showing that there was measurable contamination of glyphosate, as high as 1.8 $mg\,kg^{-1}$ of grain. The oral LD_{50} as already documented is >5000 $mg\,kg^{-1}$ of body weight. To get 5000 mg of glyphosate you would require 2.77 tonnes of soybean. Remember that this is what you require per kg of body weight. Hence if you are a 50 kg jockey or ballerina, with this upper level contamination you will need to eat 139 tonnes of soybean! You will also need to eat it raw and fairly quickly, because cooking will destroy the glyphosate and most of it will degrade within a few weeks anyway. A later study in Norway[6] detected levels of up to 3.3 $mg\,kg^{-1}$ of glyphosate in samples of soybean, meaning that the ballerina and jockey would only have to eat a mere 76 tonnes of raw soybean. I'm going to make an inflammatory statement here. That would be impossible. The largest long-distance double trailer allowable on European roads is 60 tonnes – so even one of them could not carry enough of the most contaminated soybean seed to get close to the oral LD_{50}. But there is an even bigger hurdle for you. Soybean naturally contains a strong trypsin inhibitor, meaning it prevents the normal enzymes you produce from breaking down protein in your digestive tract. Hence you cannot eat raw soybeans, or you can, but not without some damaging effects to your pancreas. Soybeans need to be processed with methodologies such as cooking so that the heat can denature the trypsin inhibitor, which in many soybeans makes up almost 2% (that's 20 000 $mg\,kg^{-1}$) of the grain.

Therefore, it is impossible to eat enough soybean to get even one-millionth of the dose needed for any toxic effects from glyphosate residue.

But what about chronic effects and what about cancer? That is the question many activists asked when faced with the reality that nobody was ever going to be poisoned by glyphosate. And it is a fair question. Carcinogens often have little or no adverse effects on individual health, yet the repeated exposure to the carcinogenic factor (whether chemical or radiation) is the risk. I'm not going to get cancer from having an X-ray but the X-ray technician who does 100 a day throughout the year is going to have problems without adequate protection.

The other class of transgenic plants was much more difficult for proponents and opponents of GM to classify. The insect-resistant maize was Bt maize. The chemical Bt was an allowed input for most certified organic farmers, and the "natural" insecticide was sprayed regularly on many fruit and vegetable crops that were classified as organic. Many scientists made the argument that Bt maize and cotton represented perfect crops for organic agriculture. The plants produced the same chemical (the Bt toxin) that was produced by the *Bacillus thuringiensis* in the insecticidal spray, but did not require the labour nor the tractor fuel to apply it, so it was, in fact, environmentally superior to the non-GM crop, especially if the non-GM version required multiple Bt applications during the growing season.

USDA was charged with developing a national standard in 1990, and it took a decade before the first regulations were put in place. This was understandable given that between 1996 and 2000, GM crop area in the USA grew to the extent that 61% of all cotton, 54% of all soybean and 25% of all maize was GM. By 2000, over 44 million hectares of agricultural land in the USA was under GM crops of some sort. By comparison, there were 490 000 ha of cropland and 225 000 ha of rangeland adhering to organic standards. The David and Goliath mindset was an obvious and appealing position to adopt for the organic industry. Goliath said GM crops are great, and organic agriculture will be so much more productive with them. David said, well we're not sure they are "natural" and on top of that, we feel uncertain about big seed companies (especially Monsanto). As the discussion went on in the late 1990s to 2000, what else was happening? Remember Flavr Savr tomatoes? Remember StarLink maize? And anti-GM activists were starting to really ramp up their activities in Europe, especially in Germany, which was the biggest European driver behind certified organic agriculture. The grandfather of organic agriculture was Rudolf Steiner, whose demonstrations of biodynamic farming in Silesia (then Germany, now Poland) started the Demeter movement in 1927.

Demeter was the Greek goddess of agriculture and the harvest, and it was Demeter who was responsible for the fertility of the soil. The Demeter movement started in Berlin in 1927, and in

1928 they took out a trademark on the name. The movement was all about producing food in concert with nature, and as well as some of the basic "values" and practices of organic agriculture, the movement was infused with a liberal dose of pagan ritual. Some of the most important activities of biodynamic agriculture include planting certain crops during the correct phase of the moon, and even more importantly the use of cow horns filled with manure. C'mon, really? I'm a scientist and I don't hold with these crazy superstitions. Except for the fact that I always (and I don't know why) have to put my left shoe on before my right shoe. That even includes when I'm buying shoes. Retail assistants always hand me the right shoe (because it's not sinister?), and I say "Can I try the left one on, my left foot is bigger". Now my left foot is in fact longer than my right foot, but it is nevertheless mostly because I just can't put the right one on first. Ahem, now back to the topic at hand. Cow horns.

Biodynamic preparation 500 involves filling a cow horn with manure, and burying it in the soil (40–60 cm deep) in autumn, then digging it up in spring. One is also required to bury a cow horn with ground quartz and following the same procedure. This is called preparation 501. Once you dig these up in spring, you are then required to make a spray preparation of both and spread $300 \, \mathrm{g \, ha^{-1}}$ of preparation 500 and $5 \, \mathrm{g \, ha^{-1}}$ of 501. Ground quartz is silicon dioxide. I don't think there is a single soil anywhere in that world that is deficient in silicon, as silicon makes up around 28% of the content of the Earth's crust. Preparation 501 fits the Hippocratic Oath perfectly: "First do no harm". Well, except if you're a cow.

Okay. Preparation 500 contains nitrogen. Nitrogen is the element most limiting to plant growth so it is an excellent idea to apply it to the soil. In saying that, preparation 500 is effectively fertilising the soil in a manner not significantly different to homeopathy. Let's assume the cow horn manure is 3% nitrogen (most well-aged cow manure is around 3% nitrogen by weight). If you follow the protocol and spread 300 g of preparation 500 per hectare, the practical outcome is that you are spreading 10 g of nitrogen (the main plant nutrient) per hectare. A small handful of urea contains about that much nitrogen. Even on good soils with nice, even rainfall, the average grain grower or vegetable grower will apply at least 100 kg of nitrogen per hectare to get an adequate yield. So applying preparation 500 is a very nice ritual, like throwing a pinch of salt over your shoulder before you eat. But it's true, it is better than homeopathy. If you're doing homeopathy properly you wouldn't be applying any nitrogen, so at least the cow horn manure contains some nitrogen, but no-where near enough to have any measurable outcome in either plant performance or product quality. It also means the plants being grown will strip considerably more nutrient from the soil than have been applied.

Let's do some simple maths. Using average maize yields and averages for nutritional content of grain, maize produces 24 g of protein per m². One hectare is 10 000 m², hence maize will produce 240 000 g, or 240 kg of protein per hectare. With the conversion factor of 6.25 (conservatively[‡]), that means that the maize harvested from an average hectare will contain $240/6.25 = 38.4$ kg of nitrogen.

So, in summary, we all have our set of values. We are all crazy to some extent. We all use logic to some extent. We will all seek to rationalise some of our crazy ideas, and some of us use science to do so. Others use a belief system that has little basis in science to explain the world they want to live in, often to the extent that it doesn't hurt anybody else and may do good. To many scientists, biodynamic agriculture is pretty crazy, pretty anti-science, and sometimes even amusing. It does nevertheless conjure up lovely, heart-warming images of beautiful rich soil laden with earthworms, pastures filled with gambolling lambs

---

[‡]There is an argument that a more accurate conversion is 5.8 in cereal grains, hence 41.4 kg N. The remainder of the biomass also has some nitrogen in it, particularly if the plant is still living and actively photosynthesising. However, let's assume that all of this nitrogen is returned to the soil organic layer post-harvest. So using preparation 500 as your sole source of nitrogen, you have added 10 g of N per hectare, and drawn 38 400 g of nitrogen out in the harvested grain. That's a big gap (38 390 g), and it is pretty hard to see how soil N depletion is going to be avoided, especially in tropical areas where the volatilisation of soil N as ammonia is more rapid because of higher soil temperatures, and more nitrate leaching occurs because of high-intensity rainfall events. Biodynamic farmers know this, or at least they find out fairly quickly, and realise that the cow horn is pagan symbolism and won't sustain a crop without a lot of other organic manures. That's why they also use preparation 502 (yarrow blossoms stuffed into bladders of deer), 503 (chamomile stuffed into cow intestines), 505 (oak bark stuffed into the skull of a domestic animal) and bury these according to prescribed rituals that help you to understand why Stonehenge made absolute sense. Nonetheless, they are producing good healthy food, so don't be too hard on them. Especially if you're one of those people who sells unsuspecting people used Citroens during the week, then goes to church every Sunday.

and heritage breed chickens, with butterflies dancing together under the sun. Imagine yourself walking through an Alpine forest listening to distant cowbells and a woodpecker looking for its lunch. You have walked 6 km, munching on a crisp apple for some energy, and then you emerge into the sunshine. Like most people you would rather emerge from the trees into a little valley with thatched farmhouses and these lovely biodynamic agrarian scenes. Then you would find a little Alpine inn with tables in the sun, and sip on a cleansing ale while pondering what to have for lunch. If you emerged from the forest to find a large metal shed filled with 50 000 hens and a McDonalds, it might rather be an affront to your values *n'est pas?*

It was Craig Cormick, the Australian science communicator, who first introduced me to this opinion driver, our "values", over a lunch in Canberra. Craig has been involved in this field of social science research since the early days of biotechnology, having worked for CSIRO and Australian government departments.

"When information is complex, people tend to make emotionally-based judgements, driven by values rather than the information presented to them. Messages that don't align with people's values or worldviews tend to be rejected or dismissed."

Craig Cormick, *The Conversation*, 9 October 2013

This is why many followers of Zen Honeycutt ("I don't need scientific studies") and Gwyneth Paltrow ("I was starting to hike up the red rocks, and honestly, it was as if I heard the rock say, 'You have the answers. You are your teacher'.") not only ignore science, but are actively against what science tells us. Sometimes, we just want the valley filled with butterflies, no matter how much science tells us many people will starve if the world turns to the restrictive practices of biodynamic farming. We can ignore the fact that the Haber–Bosch process has prevented the death of 1.5 billion people through making agricultural production more efficient. If the world went biodynamic, the demand for cow horns (preparations 500 and 501) would drive *Bos taurus* to extinction, oak trees (preparation 505) would be debarked at a rate that would lead to deforestation, and we would have to genetically modify deer to have multiple bladders (preparation 502).

And there would be that significant matter of the 1.5 billion people starving to death, plus the deaths caused by the shortage of land and water.

Like most people, I understand values, especially my own and those who agree with me. It's the values of others that sometimes require some really special efforts to treat as equal to one's own. It's a democracy (well it is where I live, which is a lucky accident of birth). Other opinions are important, and so too are values. Two politicians can use the same set of statistics to make arguments that are polar opposites. This may be based on their values. Or just as often it may be based on the fact that they have to represent their party's values, even when their own values are different.

Two of the most divisive political arguments in Australia are same-sex marriage and coal *versus* renewables. The current Prime Minister, Malcolm Turnbull, supports same-sex marriage (marriage equality if you want to put it another way) and believes in doing something sensible about preventing climate change. His predecessor, Tony Abbott, who he toppled in an internecine coup in 2015, is a pro-coal climate change denier, vehemently against marriage equality, and believes he speaks for conservative Australia. Hence the current Prime Minister has to formulate his public "values" to reflect those of the right wing of his own party. But that's politics. That doesn't happen in science. We scientists are just all too rational. We have a little sit down over a beer, discuss our science and all agree on the outcomes. Over on the next table there is a group having a little sit down over a coffee/chardonnay/kombucha, discussing the same science, and usually coming up with very similar conclusions, but sometimes not.

Surprise! Like all aspects of human endeavour (including birth, death and taxes), values and personal beliefs can and do play major roles in scientific debates. As a young scientist getting my own lab and research programme up and running, I was flabbergasted to find that some of my senior collaborators (one a virologist and the other a plant breeder) were actually creationists! These were scientists, who as part of their professional life made use of naturally-occurring and selected Darwinian mutations in everything they did, yet they had convinced themselves, because of their religious beliefs, that evolution was "not a thing".

Just look at the debate over climate change. Yes, there still remains 3% of scientists who will argue that either there is no climate change, or even if they concede that climate change is real, it's nothing to do with human activities. It is mostly their "values" that help them to retain their beliefs, and they can cherry pick some statistics and cling to the fact that these cast a shadow of doubt. It's also that 3% who regularly get invited onto the TV to cast aspersions at the consensus view, and this is an important part of the scientific debate. Realistically though, it was a lot more important when the numbers were 50:50, but start to become less relevant when they are 97:3.

Such debates are not uncommon in science. Even Albert Einstein, with all his brilliance, could not accept some accepted theories. Most notoriously he was against Heisenberg's Uncertainty Principle, famously quoted as saying, "God does not play dice with the universe". This was from a man whose very own breakthroughs in our understanding of the natural world led to quantum mechanics, which he never quite believed. As Niels Bohr once said, "An expert is one who has made all the mistakes which can be made, in a narrow field." However, these eminent scientists used scientific methodology to argue with each other. Using "values" to argue against scientific realities and clinically proven medicines is . . .. Well I'm not entirely sure what it is. One thing I do feel comfortable with saying is this. The scientific method can eventually settle arguments, and mathematics, whether as formulae or statistics can be really helpful.

- The Earth rotates around the sun.
- The Earth is a sphere.
- President Trump is orange.

These are all demonstrable truths. Yet there are people who use their "values" to argue otherwise. I will accept that the "orangeness" of President Trump can vary depending on the day, the lighting, and the colour reproduction on your television or electronic device, as well as your own colour perception (genetic variability again). And I will also accept that one day, President Trump will no longer be orange, nor will he be President. When the Earth ceases to rotate around the sun, or to be a sphere, well, you won't be reading this book so there's no point me going all Nostradamus on that one. The good news is that in a democracy, you can build your own steam-powered rocket out of scrap metal to somehow demonstrate your own "values" that the Earth is flat. Of course, doing so will very probably also improve the chances of your eligibility to receive a Darwin Award, which you will appreciate are only awarded posthumously (see darwinawards.com). Of course, some exceptional scientists have killed themselves through their scientific enthusiasm, both in the true sense (Carl Scheele who worked with many toxic chemicals; Marie Curie and Rosalind Franklin *via* radiation) and more figuratively (Lord Kelvin). Lord Kelvin, a physicist and President of the Royal Society, stated in 1902: "No aeroplane will ever be practically successful." While it could be argued that the Wright brother flights in 1903 were not all that impressive, Louis Bleriot certainly demonstrated the practicalities of flight when he took just over 36 minutes to fly from France to England in 1909, and the myriad of military practicalities of aeroplanes were clear to all by the end of 1914. Perhaps Lord Kelvin's scientific judgement was marred by his own religious beliefs and values.

Among the many studies undertaken by Craig Cormick (CSIRO study) regarding the public communication of science, he led a group of researchers surveying the public from the least to the most enthusiastic about science.[7] This led them to divide the Australian public into four almost equal segments, to which Craig gives some descriptors:

- Segment D (20% of the population) he calls the concerned and disengaged.
- Segment C (23% of the population) he calls the risk averse.
- Segment B (28%) he calls cautiously keen.
- Segment A (23%) he calls the science fans.

I'm firmly in Segment A (did you guess?). My wife and children are too. We share these values. I think my mum is in Segment B, but she tells me, "if you're doing it love, it must be good". Science fans were most likely to agree with statements like: "New technologies excite me more than they concern me" and "The benefits of science are greater than any harmful effects". The concerned and disengaged were more likely to agree with statements like: "We rely too much on science and not enough on faith" and "Science and technology creates more problems than it solves". And they drive to the airport in their hybrid cars, fly around the world to speak about it, and create blogs and post their opinions on the internet, somehow all without using science and technology?

One of the accepted theories in science communication is the "deficit model". This means that people against a particular technology or application of science have a deficit of information, and hence all that needs to be done is to correct that "deficit" with more, and perhaps better targeted, information. It hasn't been terribly successful in its application for a number of complex reasons, but one that is easy to understand. A CSIRO study that segmented the population into six segments, of which almost 40% of the Australian population fell into three categories: "Too many other issues of concern", "Science is a turn-off" or "I know enough already". These segments can be grouped as disengaged, and generally agreed with the statements "Science is out of control" and "Government funding for science should be cut". Heather Bray and Rachel Ankeny from the University of Adelaide published a study on women's attitudes towards science.[8] Interestingly, they undertook their studies with focus groups in South Australia, the only mainland state in Australia to have maintained a long-term moratorium on GM crops. The moratorium is such that if seed companies want to transport GM seed from the east coast to the west coast, or vice versa, they cannot pass through South Australia!

However, they did find that women with backgrounds in plant science or molecular biology were not particularly concerned about GM food safety, whereas women trained in health sciences and nutrition were more concerned. All women agreed that food should be natural, healthy and nutritious, local and be free of additives. Where it seemed a great dichotomy was revealed was in the attitude to risk. Some were happy that after 40 years, there was a "lack of evidence of harm", whereas the more circumspect were more concerned that they perceived a "lack of evidence of safety". Hence there was "no societal consensus about risk perception".

It's also true that there is plenty of news and social media about food. You only need to go to your favourite organic food social media feed/blogosphere to see that eating organic food is the only way you can avoid cancer/rabies/influenza. And you can get some wonderful information that you need to keep the pH of your food high (as in alkaline) to avoid cancer. Then further on in the same blog you may well see that it is important for you to increase your intake of vitamin C, oblivious to the chemical contradiction that this is ascorbic acid, which will lower the pH of your food. Then if you stay with the mainstream media you will very probably hear on Tuesday that XXX will increase your chances of getting bowel cancer, followed on Thursday by the fact that it will protect you from heart disease (where XXX = red wine, red meat, caffeine, salt, butter, oily fish, asparagus, chocolate, salami, eggs, bacon, palm oil, canola oil, bananas, seaweed, white rice, tofu, brown rice, bean sprouts, kale, whale meat, foie gras, chanterelles, white sugar, raw sugar, grasshoppers .... [insert your favourite food or drink here]). Is it any wonder we're confused?

FURTHER READING

M. S. Kaldy, Protein yield of various crops as related to protein value, Economic Botany, 26, 142–144.

REFERENCES

1. P. M. Legrain and R. Barthe, Dermite professionnelle des mains et des avantbras chez unramasseur de celeris, *Bull. Soc. Fr. Dermatol. Syphiligr.*, 1926, **33**, 662–664.
2. Contact dermatitis of celery farmers, J. G. Wiswell, *et al.*, *J. Allergy Clin. Immunol.*, 1948, **19**(6), 396–402.
3. S. F. Berkley, A. W. Hightower, R. C. Beier, D. W. Fleming, C. D. Brokopp, G. W. Ivie and C. V. Broome, Dermatitis in grocery workers associated with high natural concentrations of furanocoumarins in celery, *Ann. Intern. Med.*, 1986, **105**, 351–355.
4. L. Puig and J. M. de Moragas, Enhancement of PUVA phototoxic effects following celery ingestion: cool broth also can burn, *Arch. Dermatol.*, 1994, **130**(6), 809–810.
5. M. C. Arregui, *et al.*, Monitoring Glyphosate Residues in Transgenic Glyphosate-Resistant Soybean, *Pest Manage. Sci.*, 2004, **60**, 163–166.

6. T. Bøhn, M. Cuhra, T. Traavik, M. Sanden, J. Fagan and R. Primicerio, Compositional differences in soybeans on the market: Glyphosate accumulates in Roundup Ready GM soybeans, *Food Chem.*, 2014, 153.
7. C. Cormick and L. Malzoni Romanach, Segmentation studies provide insights to better understanding attitudes towards science and technology, *Trends Biotechnol.*, 2014, **32**(3), 114–116.
8. H. J. Bray and R. A. Ankeny, Not just about "the science": science education and attitudes to genetically modified foods among women in Australia, *New Genet. Soc.*, 2017, **36**(1), 1–21.

CHAPTER 9

O Fortuna!

"Come what may, all bad fortune is to be conquered by endurance."

Virgil, *Aeneid*

BASF, "The Chemical Company", has its base in what can most politely be described as an industrial part of Germany straddling the river Rhine. In 2011 I was visiting Limburgerhof, near Ludwigshafen, the worldwide headquarters of BASF. Limburgerhof is the agrochemical centre of BASF, a small place very much dwarfed by the industrial might of Ludwigshafen on the west bank of the Rhine in Rhineland-Palatinate. Across the river is the even larger and perhaps even more industrial city of Mannheim.

Arriving by train from Freiburg and the beautiful Black Forest, there can hardly be a greater contrast. Medieval towns, babbling brooks, quaint little churches and hills covered in autumn leaves, gave way to a totally different landscape in less than two hours. The Rhine valley of vineyards was picturesque enough, but soon the train passed into a flatter landscape and the chimneys of Mannheim and Ludwigshafen dominated the horizon from quite some way away. As you draw closer, not just the smoke is visible, but some of them appear to be emitting flames, like modern-day dragons.

Good Enough to Eat? Next Generation GM Crops
By Ian D. Godwin
© Ian D. Godwin 2019
Published by the Royal Society of Chemistry, www.rsc.org

The area around Ludwigshafen has been conquered by many over the last 2000 years, being under Roman control at that time, and having endured various times under French rule. Mannheim, over the river has a longer history, and in truth, Ludwigshafen really only took off as a city when the people of Mannheim told BASF that they should put their smoky, smelly air-polluting dragon-breathing factories over the river. In 1865, BASF moved their operations to the Ludwigshafen side of the river, where amongst all those lucrative blue dyes that made their name, they produced poison gas, first used in World War One. One of the outcomes was the historic distinction of Ludwigshafen becoming the first target of a strategic bombing raid. In 1915, French bombers specifically targeted the BASF factory. In the resulting terror, 12 factory workers were killed. This later paled into relative insignificance. During World War Two, Ludwigshafen was bombed in 121 separate raids, and as a result there was not much of the city left by March 1945, contributing further to its utter paucity of civic beauty. Mannheim across the river is still in Baden-Württemberg, but it's not exactly the Black Forest either. Also extensively bombed in the early 1940s, it does have a few reconstructed historic buildings, yet probably not enough to excite anybody other than an industrial historian. It was in the streets of Mannheim in 1886 that a fellow called Karl Benz drove the first automobile with an internal combustion engine. Bizarrely (or I think so), I have seen one of these first models, looking like a carriage without a horse. The one I saw was in a small town called Balcarce in Argentina. How did it get there? Well a certain Grand Prix driver called Fangio has a museum in his honour in his home town, Balcarce. Given that Juan Miguel Fangio won no fewer than five world championships, and two of them with Mercedes-Benz (1954–55), when he retired he became the President of Mercedes-Benz in Argentina. Balcarce is also a very good place to eat steak and potatoes, so let's get back to the potatoes.

I was meeting members of the BASF crop biotech team at the Limburgerhof Research Station. They were cautiously excited about their new soon-to-be released potato, Fortuna. Fortuna was named after the Roman goddess of agriculture, and was of course a lady of plenty and fecundity. There was muted excitement about the new potato. The excitement was tangible because BASF and their partners were confident that this was the cleanest, greenest way to deliver potatoes free of the scourge of late blight. It may even be an attractive option to organic farmers so they didn't have to contaminate their land with toxic copper and accept yields 50% lower than conventional. BASF was hoping this could become their "once more unto the breach" moment.

Now set the teeth and stretch the nostril wide;
Hold hard the breath and bend up every spirit
To his full height. On, on, you noblest ... scientists
After William Shakespeare, *Henry* V

The excitement was muted because:

1. The potato had been developed much earlier and had taken almost a decade to get approval for release.
2. Activists were ready and waiting, battleaxes and swords drawn, still green with the sap of all the maize, wheat, potato and other crops they had destroyed in the name of their campaign for genetic purity or whatever it was. They were on a roll, and had been confidently destroying GM crops all over Europe. Well, and quite a few non-GM crops too, but collateral damage is inevitable in war, right?
3. It was Germany. Remember "80% of the population are against GM crops and the other 20% don't care".

Late blight of potatoes is caused by one of the most incredible and difficult plant pathogens known to us, *Phytophthora infestans,* as already discussed in Chapter 3. *Phytophthora infestans* has a very complex genome,[1] and at 240 million base pairs of DNA is larger than some plants, such as *Arabidopsis*, the model plant with a genome around half that size.

The disease arose in Mexico,[2] and made its way to Europe several times on infected tubers which were used as planting materials. The disease was caused by an oomycete – a fungal-like disease, but not really a fungus. While most plant diseases have an impact on productivity, at least if you can harvest some yield from the infected plant, you at least have something to eat. As an insidious disease, late blight leads to not only the death of badly infected plants, but also can quickly render the potato inedible. Even at a moderate level of infection, the tuber becomes darkened in colour, and will usually rot quite quickly, hence the ability to store potatoes throughout the colder months is lost.

It was first recorded in Europe in 1843, and by 1845 had led to the Irish Potato Famine and the Highland Potato Famine in Scotland. The Irish famine was particularly devastating because almost the entire food supply was based on one highly susceptible cultivar, the Irish Lumper. Here was a disease which almost singlehandedly contributed to over 1 million Irish deaths and the fact that almost 2 million Irish citizens emigrated to places like the USA, Canada and Australia. It can truly be said that the Irish population has never recovered (Figure 9.1). In the Scottish Highlands, the populations at risk were considerably smaller, but a Scottish diaspora was also an outcome.

I have used the term "almost singlehandedly" because to make the famines anywhere near true disasters there is very often a requirement for political interference and market distortion. The mainly English landlords in Ireland were still exporting grain to England at inflated prices which the Irish could not afford throughout the terrible shortage of potatoes. The Highland Clearances in the Scottish Highlands had been going on since the previous century, and the potato famine enabled the final push to de-populate the area (of tenant farmers) to raise sheep to be expedited and completed.[3] The disease also arrived

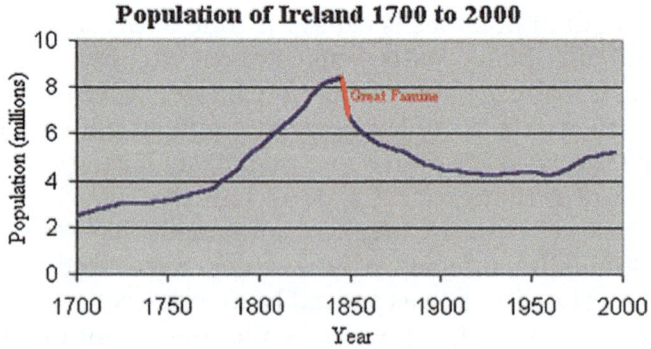

Figure 9.1   Population of Ireland from 1700 to 2000.
Image by Wesley Johnston (http://www.wesleyjohnston.com/users/ ireland/past/famine/demographics_pre.html).

throughout continental Europe in the 1840s and while not causing famine on the scale of that experienced in Ireland, did contribute to serious food shortages and social unrest, as many places in Europe experienced severe food shortages in 1846. Many historians believe that this hardship amongst the rural and urban poor probably contributed significantly to some of the revolutions seen across Europe in 1848, most markedly in France, Germany and the Netherlands, where potatoes had become an important staple. You could say that the unrest of the 1840s and beyond started with a powerful combination of a tiny little oomycete from Mexico (bloody immigrants!) and the meeting of two rather larger bearded gentlemen called Karl Marx and Friedrich Engels in 1844.

Unfortunately, blight was a disease that didn't go away. It was always there as a silent and hopefully unseen threat, rearing its head when wetter and cooler springs/summers were experienced. However, it was soon determined that spraying the crop with copper was a reliable protection against the disease. The most effective forms were copper sulphate or Bordeaux mixture (copper sulphate plus slaked lime), which have the action of preventing the germination of spores. Hence these have to be used prophylactically, and are applied more frequently during wetter weather.

Disaster struck again during World War One in Germany. Weather was perfect for the outbreak of late blight on the potato crop of 1916, and there was no copper. Germany was suffering from an Allied blockade and all available copper was being used in munitions and for electrical wire. No copper meant no protection against late blight and the potato crop failed badly. Grain was already in shortage, and the war effort was already substituting potato bread for wheat and rye. Now the German government turned to turnips for their salvation. To the German populace, 1916–17 went down in history as the Turnip Winter. Female mortality skyrocketed (because the males were on higher rations due to military service and war effort labour), youth death rates increased, and youth crime statistics went through the roof as young people tried to find a way to feed themselves. In 1917 the Navy revolted over the quality of the food, and by 1918 it is estimated that over 700 000 people, mostly civilians or wounded soldiers (also on lower rations) had perished due to starvation and malnutrition.[4] Copper continued to be the control measure of choice until safer fungicides were developed.

Then something just as revolutionary happened in the 1980s, and it was another revolution of Mexican origin. Other oomycete spores made it to Europe, and these ones were just as unwelcome as the ones that arrived 140 years earlier. The 1840s arrivals had been living a celibate life, and reproducing asexually. We now know this as mating type A1. The new oomycete population consisted of both A1 and A2 mating type, and when it arrived in Europe in 1981, the sexual shackles were cast off.

Are you thinking what I'm thinking, A2? I sure am, A1! Fungal sex was back on the menu, and with this newfound sexual freedom came the biggest driver of genetic variation we know of – genetic recombination! The outcomes for the potato world were potentially disastrous. Through decades of potato breeding, new varieties with resistance to late blight had been developed. The pathogen was slowly overcoming these resistance sources *via* mutation, but clever plant breeding combined with the use of fungicides was keeping potato production ahead of the speed of disease evolution (Charles Darwin would have loved it!). In 1981, the new mating type was demonstrated to have been found in Belgium, Switzerland, the UK, the Netherlands and East Germany. A group at Wageningen University[5] in the Netherlands traced the spread of A2 and within four years it had been found in the Soviet Union and as far east as Japan, as well as the Middle East.

Once A1 and A2 coexist and can reproduce sexually, the disease can form something called an oospore, which is much tougher and longer-lived than the asexual spores. When only the A1 type existed there were only four races (known because they could attack one of four known resistance genes) and they could all be controlled with metalaxyl. By the early 1990s, there were 73 different races identified, and that was just from surveys in the Netherlands,[5] and many were resistant to the main control, metalaxyl. Hence, since the early 1980s, potato production in Europe and beyond has become biological and chemical warfare.

Fortunately for me, André Drenth is a colleague and Professor in Plant Pathology at the University of Queensland. Over coffee one sunny Brisbane morning, André told me how he came to be part of the group to determine that new A1 and A2 strains had well and truly arrived in Europe. André was brought up on a potato farm in the province of Groningen in the northern part of the Netherlands. His family farmed land that until 1914 had been peatland. As the Dutch have demonstrated for centuries, they have developed particular skills at draining and reclaiming land, even if it was originally below sea level. Farming in this area on newly developed peatland is very challenging as these are very organic soils, with lots of micronutrient problems, which was seriously limiting to the achievement of good yields. "Organic" in this case means there is not much mineral content to the soil, such as useful things like clay and sand. They are usually poorly drained, and commonly quite acidic. Still, they are "organic" in the true sense so they must be good, right? Paradoxically, organic soils, also known as histosols, have very low fertility and are very commonly almost totally lacking in micronutrients such as boron, copper and zinc.

His grandfather was good enough to be one of the two farmers of the 13 nearby that did not go broke during the 1930s, when the Great Depression meant that the crop was worth very little money. Then he also survived the 1940s when the occupying Germans appropriated crops and equipment, while fuel and other inputs were near impossible to obtain. After a brief period at college to attempt to become an engineer, André went back to help run the family farm, before going to a junior agricultural college where he took the opportunity to travel the world ("bummed around while trying to find out what I wanted to do") working on farms in Texas, Kenya and France. Then he decided he wanted to study agricultural science, and more specifically plant breeding at Wageningen. In his fourth year he spent 8 months with the famous wheat rust pathologist, Bob McIntosh, at the University of Sydney. He enjoyed it so much he decided to study for his doctorate, and was interested in something that combined plant genetics and phytopathology.

He was offered a position to study the genetics of late blight, a disease he was familiar with on his own farm. Initially he was very reluctant. "There had been over 140 years of scientific research on late blight and potatoes – what else could there possibly be left to discover for a new PhD student?" Within the first year of his study, he had the answer to the question. Using DNA markers, André was able to show that something had changed genetically in isolates of late blight collected before the 1980s and those collected since. Sometime in the late 1970s, it seems new A1 and A2 mating type isolates of the fungus had arrived in Europe. He believes that the evidence suggests this happened around 1976, when much of northern Europe was experiencing an historic drought. The drought led to a potato shortage, and hence potatoes were being imported from North America. It was forbidden to import potatoes from Mexico, yet it seems this is exactly what happened. Potatoes from the Toluca Valley in Mexico, the epicentre of late blight diversity, landed somewhere in Europe, probably accompanied by dodgy paperwork claiming these potatoes were citizen tubers of another country. Using DNA evidence, it was definitively demonstrated that the new A1 and A2 strains had come from Mexico.[6]

The genetics of the original A1 strain of *Phytophthora infestans* was that it is a tetraploid, meaning it has four copies of each of its chromosomes. The new A1 and A2 strains were diploid (only two copies of the same chromosome) and André demonstrated by combining dirty boots field research with molecular genetics that the sexual oospores were formed and that recombination was happening in potato fields.

The outcomes were devastating. The sexual cycle meant:

- The floodgates of sexual diversity were opened.
- The diploid version of the disease aggressively infected potato, and it was more devastating because it led to stem collapse and plant death.
- Being an oomycete, it could spread extremely quickly, as had already been seen with the first arrival back in the 1840s.

As André told me, late blight first arrived in the USA from Mexico in 1843. Then with transatlantic trade, the disease was first spotted in Belgium in June 1845. By mid-August, it was throughout the Netherlands, and had incursions into the south of England, the west of Germany and northern France. By October, it had made it as far south as Spain, had gone west to cover the whole of Ireland, north to Denmark, Norway and Sweden and was as far east as Poland.

The Atlantic Ocean was a barrier to the spread of the disease, yet humanity intervened and took the disease on infected potatoes to a monastery in Belgium. The disease could then do the rest of the pestilent work itself. Bodies of water such as the English Channel, the Irish Sea and the Baltic were no barrier to spread. In the moist, cool environments of northern Europe, millions of spores could be produced on a single leaf, and these could spread on infected potatoes, in water, and on the wind. Spores could spread many, many kilometres. By the late 1850s the disease was documented across much of Russia and even as far east as China, and south throughout Africa. As described, the effects were far-reaching, but at least most people had other sources of carbohydrate such as wheat and rye. Not so the Irish, because the wheat was all being shipped to England. As tradition had it, farmers saved a proportion of tubers for next year's plant crop. Blight rotted many of these stored potatoes over winter, and those that survived carried the disease into next year's crop. By 1847, there was a serious shortage of potatoes to plant. The second wave of devastation, courtesy of the new A1 and A2 strains, followed a similar pattern. Having arrived in Europe in the 1970s the disease was everywhere in Europe in the 1980s and from there, spread south through Africa and east throughout Asia.

One of the few places the disease had not troubled was the tropical island of Papua New Guinea. Potatoes? In the tropics? Yes, indeed. In the Papua New Guinea Highlands you can grow almost anything, and significant amounts of potatoes are produced in the mountain regions around Mt Hagen and Goroka. However, the trans-migration policies of the Indonesian government meant considerable numbers of Javanese were moving to the western half of the island, the Indonesian province of Papua. With them came potatoes for planting, and early in the new millennium, it happened that some of the new arrivals (the potatoes, not the Javanese) carried late blight. The inevitable had arrived, and here was a modern Highlands potato famine about to unfold. Looking around for a nearby expert, the call went out to Professor Drenth for help. The PNG situation shared some striking similarities with Ireland of the 1840s. The climate was very wet, and because of the altitude of the Bismarck Ranges, relatively cool for the tropics, with night temperatures a remarkably stable 11–12 °C all year, and 22–26 rainy days every month. To make things worse, almost all the potatoes grown were cv Sequoia, a US-bred variety originally released in 1940. Sequoia is highly susceptible to late blight, and when the disease was first detected in the western Highlands in January 2003, it quickly spread to Mt Hagen and then east to Goroka by April, an average of 60 km per month. Fortunately for the Highlands, the most densely populated part of PNG, there was also plenty of sweet potato to eat, or the results may have been devastating. Like the circumstances around the second incursion into Europe, the worst drought on record across the island in 1997 led to the importation of potatoes to overcome food shortages and shortages of seed potatoes. These potatoes were probably carrying the disease.

Copper remains a useful prophylactic, yet it cannot overcome the disease once it has infected the plant. This means, among other things, that copper has to be applied quite frequently as new and thus unprotected leaves are formed continuously, while during wet weather copper washes off and spores of the disease are more prevalent. Copper sulphate (which any chemist knows is inorganic) remains the main weapon for organic potato growers. As André Drenth pointed out to me, Bordeaux mixture was not actually invented as a means to control fungal infection, it was invented as a means to deter thieves. Grape growers in Bordeaux made a mixture of copper sulphate and lime to spray along the vines near the roadside. It was highly visible and created enough of a bitter taste to discourage passers-by from helping themselves to the grapes. Pierre Millardet, a professor of botany at the University of Bordeaux made the observation that there was very little mildew in the grapes beside the road, whereas the rest of the vineyard was badly affected. He then undertook some trials of the mixture at Chateau Dauzac in the Margaux region of Bordeaux, where, with the help of the vineyard's technical director, Ernest David, he was able to demonstrate the efficacy of the solution against downy mildew infection (Figure 9.2). For this reason, the mixture is also known locally as the Millardet–David treatment.

In the 1920s and 1930s, the United Fruit Company (now Chiquita) used large amounts of copper sulphate and Bordeaux mixture in their production of fruit and vegetables in Latin America, especially to control diseases of banana. The farm workers in Puerto Rico called the workers who sprayed the

Figure 9.2 Downy mildew (*Plasmopara viticola*) fungus on grapes – the primary need for fungicides in vineyards.
© Starover Sibiriak/Shutterstock.

copper chemicals "parakeets". The term was not one of fondness, but was used to describe the fact that the workers who applied it without modern protective equipment would soon turn blue-green, and many became ill (respiratory problems, infertility, weight loss, cancer to name a few) and died from copper toxicity.[7] There were some reports of similar symptoms among vineyard workers in Europe, yet copper sulphate and Bordeaux mixture are still in widespread use.

Conventional potato growers, who are able to use modern and safer control mechanisms such as fungicides, achieve twice the yield on average when compared to organic or pagan (OK they call themselves biodynamic) potato growers. What sacrilege am I spouting? A fungicide is safer than an organic allowed input? I should wash my mouth out with rotenone!

Remember those good old statistics and in particular the oral LD_{50}? The main fungicides used to control late blight are mancozeb, cymoxanil and propamocarb. Here are some figures (Table 9.1) comparing these to the copper-containing compounds – I know, yet again they are in rats. Well that's because there are (I know, really?) a few ethical issues when it comes to

Table 9.1 Toxicity of major chemicals used to control potato late blight.

Chemical	Oral LD_{50} rats (mg kg^{-1})	Other notes on toxicity or environment
Mancozeb	>5000	Very toxic to aquatic life
Cymoxanil	No toxicity, poss. chronic	Toxicity to birds >2000 mg kg^{-1}
Propamocarb	>5000	
Bordeaux	2174	
$CuSO_4$	482	Very toxic to aquatic life with long-lasting effects

working out how much of a given substance it takes to poison the "average human". Unless, almost inexplicably, you are anti-glyphosate anti-GMO campaigner Séralini who has done taste tests of pesticides in water and wine.[8] Many wonder about his human ethics clearance. Many also wonder about his choice of journals – such as this one recently published in a journal called *Food and Nutrition Journal* by Gavin Publishers. Other than saying they are listed on a number of sites as predatory publishers, this study is a wonderful demonstration of what you can get published if you're willing to pay. By way of full disclosure, I haven't tasted any of the fungicides listed below, but if I had to, I think I would select cymoxanil to go with my fries. Yet as you can see, copper sulphate, the inorganic darling of the organic farmer, is at least ten times more toxic than the modern fungicides (which are organic compounds).

Copper is also very persistent in the soil and accumulates to reach levels where it becomes toxic to worms and other soil organisms. Being inorganic it sticks around basically forever. Slow increase of copper in soil is a problem, particularly in vineyards where it has been used for over a century.

André Drenth is of the opinion that the only solution to potato late blight is genetic. He told me that there has been tension between conventional and organic potato growers in Europe. The conventional producers, who have better chemical control regimes, blame the organic growers for the build-up of inoculum. He says that humans and genetics have never been the easiest of combinations. Genetics can be complex, and after some of the eugenics experiments around the world in the 20th century, he feels that somehow genetics are not regarded as "natural". There are many wild potato species, yet all of them have small tubers and produce toxic amounts of bitter-tasting alkaloids. He ponders whether it may actually be possible that the first people to domesticate potatoes had to deal with these toxins by post-harvest treatments like what is still done with cassava to remove cyanogen.

André told me that during pre-Colombian times in the Andean Highlands fresh wild potatoes were cooked in an oven and dipped in a sauce made of clay, water and salt prior to eating. We now understand that this was done to absorb and thus detoxify the glycoalkaloids present in wild potatoes. Luckily our modern potatoes have very low levels of these toxins so we can eat our fries with ketchup instead of a clay sauce. Another technique used to enable safe eating of wild potatoes was freeze-drying. Believe it or not, freeze-drying has a long history in food preparation. This method was a traditional means of processing and storing potatoes by the Andean peoples of Peru and Ecuador. The process produces chuño, which involves spreading the potatoes out in freezing temperatures overnight (at almost 4000 m altitude) and spending the days in the sun so that they dehydrate. Once the process is over, the freeze-dried potatoes are free of toxins and may be stored for decades.

What really interests geneticists is that these wild potato species carry a huge genetic resource, in that they have many genes for late blight resistance. Unfortunately, attempts to cross these with cultivated potatoes are difficult and often unsuccessful, and hence the only way to transfer these naturally-occurring resistance genes into potato is *via* genetic engineering. This means that even though the resultant potatoes contain nothing other than one or two genes from a wild potato, most jurisdictions regard them as GMOs and the activists can mount their environmentally unsustainable arguments against them. So chemical sprays are the only option. Not the only option. Not growing potatoes is another option. That's not an outcome André (the potato farmer and the geneticist) wants to consider. Not only does he love his potatoes, he is adamant that they are one of the best food crops available to us. As he points out, with cereals, we harvest about one-third of the biomass. With potatoes, we harvest two-thirds of the biomass, and they have a decent protein content, as well as vitamin C. You can almost subsist entirely on potatoes. And, you can make fries.

The Fortuna potato had two great features. The first feature was that it was a good, high-yielding multipurpose potato based on a variety called Fontane, and suitable for making fries. The second feature was that it was effectively Fontane with two new genes, and they were both from wild potato relatives. These two genes, befitting for such an awesome outcome, are excitingly called Rest in Peace – Blight Basher (RIP-BLB1) and Rest in Peace – Blight Bonker (RIP-BLB2). No actually, I just made that up. But that's what they would have been called if a geneticist named them. They were, in fact, named by plant pathologists so they were called Rpi-blb1 and Rpi-blb2. Geneticists come up with gene names like Sonic Hedgehog (that's to do with the space between your eyes), INDY (the I'm Not Dead Yet gene which doubles the lifespan of a fruit fly) and Ken and Barbie (which produces fruit flies with no external genitalia). Plant pathologists like to give names like Lr1, Sr34, Rp-1D, C3-PO, *etc.* Plant pathologists are some of my favourite people. No really, they are. The source of these blight resistance genes was a Mexican wild relative, *Solanum bulbocastum*, known as the ornamental nightshade. Plant breeders had been unsuccessful in getting this nightshade to cross with cultivated potato, hence the only possible approach was to clone the genes and transform them into a potato. The ornamental nightshade evolved in the presence of the A1 and A2 mating types of *Phytophthora infestans* and has resistance to all known strains, which made it the ideal donor of disease resistance.

BASF applied to get the Fortuna potato released in 2011. It was resistant to all known strains of potato late blight, a disease that was still causing somewhere in the order of €2.4 billion or £3.5 billion (and lots of other figures in between – it depends who you ask) worth of damage to the potato industry worldwide each and every year. What we do seem to agree on is that world potato production is about 20% lower on average because of this disease. Potato crops often get sprayed 10–15 times per year and even more in a wet year (conventional and organic). That is costly, takes a lot of labour, and results in a quite substantial carbon footprint from all that diesel. So this is a situation just like the cotton and Heliothis problem, except it's on a grander scale and potato is one of the four most important food crops for people everywhere, not just at McDonalds.

In a shock announcement in early 2012, BASF announced it was withdrawing the application, citing consumer and political resistance. Furthermore, they were halting their plant biotechnology research in Europe and moving BASF Plant Science to the North Carolina Research Triangle in the USA, with the loss of 140 jobs in Limburgerhof. The company that once had to move its factories across the Rhine, which must have been no more than 300 metres, was now on the move again. This time it was across the Atlantic, almost 7000 km away. This was a big deal. A German company giving up on Germany and moving operations across the Atlantic. There were some definite attractions, given that there are three world-class universities (North Carolina State, University of Carolina and Duke), as well as a large number of biotech companies in the Triangle (GlaxoSmithKline, Biogen, Merck, Novo Nordisk, Novozymes, Pfizer, DuPont, Syngenta and Bayer).

Friends of the Earth were popping the champagne open – hopefully organic:

> *"This is another nail in the coffin for GM foods in Europe. This is a good day for consumers and farmers and opens the door for the European Union to shift Europe to greener and more publicly acceptable farming."*
> Adrian Bebb, FoE, cited in *Chemistry World*, 18 January 2012

Not surprisingly, I don't see the cause for celebration. I get the greener part, because he was probably meaning the parakeet blue-green that copper was causing over the farms of organically farmed potatoes. But greener in the environmental sense? Here was a genetically resistant potato, and it was going to save up to 25 sprays per season in a bad year, which amounts to huge costs, lots of diesel fuel and I can't see any environmental benefit from this at all, not even accounting for the copper.

It was hailed by activists all over Europe that this was the beginning of the end and the death knell was chiming for GM crops in Europe, and then, the rest of the world too. Fortress Europe had defeated the evils of GM crops, and the future was organic. BASF had left the building to join Bayer CropScience in North Carolina. In 2014, KWS, the seed company in Einbeck in northern Germany, established a research base in St Louis, not far from Monsanto.

The activists had won. GM crops in Europe were tod, mort, stuffed, muerto, kaput, død, kuollut, morto, marbh, nekros. All that was needed now was for everyone to go biodynamic. The purveyors of the finest yarrow, cow horn and deer bladder were going to become the *nouveau riche*. Chile was going to be a huge beneficiary. The world price of copper was going to go absolutely ballistic, and Chile digs more of it out of the ground than anywhere else in the world.

Dennis Eriksson is a Swedish molecular biologist who has turned his hand to understanding and informing European policy on crop genetic manipulation. His scientific work has been on the development of better oils from crop relatives of canola. One frosty morning in November 2017 I visited Dennis at the southern campus of SLU (Swedish Agricultural University) in Alnarp, not far from Malmö. Like many historic university campuses in Europe, it has a castle on campus. Naturally, this is where the administrators are, not the scientists. Dennis has been working with other scientists across Europe to attempt to deliver some consensus to the debate across the EU. He told me:

> *"The EU is good for many things, but when it comes to GMOs it somewhat tricky. When you have 28 countries voting on every item of legislation, there has never been any clear majority for or against the cultivation of GM crops. Note that is for cultivation. With 70% of the protein for animal feed imported, mostly as GM soy, the commission has had to accept GM crops for food use. Just so long as they are not grown in Europe."*

He told me that there is currently only one GM crop grown in Europe, the Bt maize with MON810, and only about 120 000 hectares are grown in Spain, Portugal and the Czech Republic. There has been and still is lots of opposition from green NGOs, and although the European Commission would like to act to accept more GM crops, it has never dared because of the pressure from NGOs. He spent some time in Brussels, and told me "in reality the Commission are not the problem, it's the Parliament". Some members of the European Parliament are themselves activists, such as the French member, José Bové. A former sheep farmer, Bové has made himself a career out of protesting against McDonalds and meat containing hormones.

He took on the World Trade Organization, and by association this led him and his followers to destroy a half-built McDonalds restaurant in Millau in southern France. With his group, he participated in the destruction of a number of GM field trials, and famously in 2008, went on a hunger strike against the cultivation of GM maize in France. Bové joined the Greens (Europe Écologie as they are known in France), then was voted into the European Parliament in 2009. Bové is not one to stand back and listen to others. Most of his activities as an MEP have been on the Committee for Agriculture and Rural Development. With other Green MEPs, he has blocked numerous recommendations (from the European Commission) to allow the release of GM crops for cultivation. The debate still rages, and unfortunately it is not about the science (*quelle surprise!*), which – whoever measures it – shows the overwhelming benefits of the cultivation of GM crops. The benefits are social, economic and environmental. Sadly, in the country where the first genetically modified plants were made at the University of Ghent, the MEPs in Brussels still manage to inculcate fear and loathing when it comes to the cultivation of GM crops. On a different, and yet related matter, France now has almost 1500 McDonalds restaurants. In Europe, only Germany has more McDonalds outlets, but it does have a significantly higher population. When expressed in number of McDonalds restaurants per million people, there are three countries in Europe with around 20 outlets per million. Those countries are Austria, Sweden and France.

While in Brussels on a grant from the European Plant Science Organization, Dennis Eriksson gained first-hand knowledge of "how sensible the commission is, and how crazy the system for release of a GM crop". He detailed to me the case of the Amflora potato, another BASF product and only the second instance of a crop to be granted approval for cultivation in the EU. Amflora potatoes were designed for the production of industrial starch. Downregulation of a gene known as granule-bound starch synthase meant that the potato would only produce amylopectin and little or no amylose in its starch. Amflora was not designed to enter the human food chain, and was aimed at the European starch market, which is predominantly met by potatoes and wheat. Industrial starch is used for coating paper, cotton and other fibres, as well as an additive to concrete and other industrial polymers. BASF produced the Amflora potato in the early 1990s and performed many of their field trials in Sweden. In 1996, BASF submitted an application for cultivation in the EU through the Board of Agriculture in Sweden. Unfortunately in 1998 the application hit a five-year moratorium on the release of GM crops for cultivation. However, there was an acceptance that patience was needed. Once the application went to the European Commission, it was then required to go to all Member States. If no Member States object, a hurdle completed in 2007, the application can then proceed to the next step, which is for the European Food Safety Authority (EFSA) to assess the crop for any potential risks to human and animal health and the environment. Under the legislation, EFSA is supposed to complete this in six months. A study by Europabio found that EFSA takes four to five years on average to release its decision, predominantly because of the political pressure brought to bear.

Finally in March 2010, Amflora was deemed to be safe and was released for cultivation for industrial use in the EU. BASF then proceeded to plant the potato in Germany and Sweden for the production of more planting material, and in the Czech Republic for commercial production. Naturally the Greens Alliance and other environmental groups protested, and some member countries, including Hungary and France, commenced a legal challenge to the approval. Approval had taken 14 years to get through the cumbersome European system. Within two short years, BASF announced that it would no longer cultivate the potato in Europe, and move operations to North America. In 2013, the EU revoked its approval for cultivation. As you read earlier in the chapter, the revolutionary clean and green, late blight-resistant Fortuna potato soon followed. The anti-GM activists rejoiced, as did the manufacturers of copper sulphate (organic potatoes) and modern, safer, less toxic fungicides (conventional potatoes). Pelted with rotten potatoes, over-ripe tomatoes and mycotoxin-infested maize, GM crops exited stage left, without even bowing. It was all over for GM crops in Europe.

Then came the Oxford Farming Conference in 2013. A man in his 30s came to speak at the podium. He wasn't a farmer. He wasn't a scientist. He wasn't an agro-politician. He did not work for an evil multinational agrochemical or seed company. He was a man who had experienced the clear and present danger of GM crop plants for many years. What he said both amazed and shocked the audience:

> *"My lords, ladies and gentlemen. I want to start with some apologies, which I believe are most appropriate to this audience. For the record, here and upfront, I apologise for having spent several years ripping up GM crops. I'm also sorry that I helped start the anti-GM movement back in the mid-1990s and that thereby I assisted in demonising an important technological option which can be used to benefit the environment. As an environmentalist, and someone who believes that everyone in the world has a right to a healthy and nutritious diet of their choosing, I could not have chosen a more counter-productive path. I now regret it completely."*
>
> Mark Lynas, 2013

Mark Lynas, the eco-warrior and anti-Monsanto activist who wrote regular anti-GM crop pieces for *The Guardian* and appeared on various television shows, had not so much experienced a change of heart, but a change of opinion. You could say he had decided to get a more informed opinion. He actually decided he needed to understand how GM crops work and what their applications are, and once he did the research, his informed opinion involved understanding the science. One of his first books, *High Tide*, was all about the role humanity was playing in climate change. Delving into the science of climate change, which was a departure from his background education in history and politics at Edinburgh, he soon came to realise that he could not accept and embrace the scientific consensus around climate change, yet at the same time reject that same level of scientific consensus about the safety and environmental benefits of GMOs.

He had once blindly accepted Greenpeace's opposition to GM crops, but now he has really taken issue with the fact that Greenpeace and other organised activist groups have slowed the release of the high beta-carotene Golden Rice project. He has recently gone so far as to write a book published in 2018. *Seeds of Science: Why We Got it So Wrong on GMOs* (Bloomsbury Publishing) will be Mark's take on how public opinion has been turned against GM crops and how this misplaced opinion flies in the face of the science (and the economic, social and environmental benefits). As I write this book, I look forward to the release date with great anticipation (after I rid myself of the disappointment that Mark Lynas, The Mark Lynas, the one who won the Royal Society Prize for Science Writing in 2008, and gave a world-famous speech in 2013 is publishing a book on GM crops in the same year as me. Crikey!). As an academic, I am always encouraging my lab members to write "active" titles for their papers:

- Passive title: The effect of lead on the human brain
- Active title: Lead causes brain damage in humans

The active title tells the story, piques the interest. Maybe I should rethink my book title? But then again, I have always enjoyed reading mysteries. And this one is a mystery, except so far there has not been a death attributed to GM crops and foods.

In late 2017 I had the great fortune to speak to Mark in London while I was in Denmark. The interview started with a huge amount of background noise and he had to move upstairs to escape those family virtues of children getting ready for school (you may know that noise, the pandemonium, shouting and clanking that can resemble that of a small Viking melee). Finally locked into his study, we could actually hear one another. I asked about the weather (he is English), and then I asked about Mark's upbringing. He told me his dad was a geologist who worked for the UK government, and part of Mark's early life was spent in exotic locations like Peru and Fiji. His parents "retired" to North Wales where they started to farm, and were very attracted to organic agricultural principles. However, Mark's description of the farm is that it is post-organic (meaning it is no longer certified organic). His father wants to sometimes use "a little bit of glyphosate to control weeds" among other "non-allowed inputs". He regards the constraints imposed by organic farming as pointless, because they do not lead to better environmental outcomes nor are they supported by science. One of the more frequent complaints from his father is along the lines of "Why can't I use blight-resistant potatoes if I want to?" He sees the use of these advanced technologies as environmentally beneficial and more productive (which, frankly, they are!).

Mark's research into the science of GMOs convinced him that this is the case too.

I asked Mark if he still regards himself as an eco-warrior. He told me that being an eco-warrior makes life fun and meaningful, and at first, sneaking around in the dark to hack down GM crops with a machete was quite fulfilling. He told me he is still a staunch environmentalist, and he still wants to save the world, however, "now I am quite a fan of empiricism and rationalism, and one can harness those to save the world". He is actually involved in the Alliance for Science network, currently headed by Sarah Evanega at Cornell, and is helping out with the release of Golden Rice. He says Greenpeace cannot allow Golden Rice to be successful, because it will expose the fact that they have been wrong and lying about it for so many years. He has already revealed in past interviews many times that he has lost many good friends, including his best man, since he changed his opinion about GM crops. Much of it comes back to values, and he has been quoted as saying "you've insulted people at the deepest level of their values" (*The Guardian*, 2013). He was also fairly forthright in his opinions of some of the more high-profile eco-warriors, such as the "physicist" Vandana Shiva. "She's just a lunatic, and the sorts of things she says make her look quite loony. She's not a great fan of mine and I think you can summarise to say the feeling is mutual. And by the way, she's not a physicist. Her PhD was on the philosophy of physics."

We spoke of the regulation surrounding GMOs. He is of the opinion that it is not quite so straightforward as to say that GMOs are the same as the outputs of plant breeding. For a start, there is a spiritual and ethical dimension. Taking a gene from one organism and putting it into another organism is quite different in most people's minds to what happens naturally in a forest. Now once you add the spiritual or ethical dimensions to the argument, this has the propensity to lead to deeper concerns. If you find the concept of genes going from one organism to another somewhat repulsive, then naturally you are going to question the health and safety of the new organism.

Then if you are already of an ideological bent and think that large multinational corporations are inherently evil, or at the very least, not particularly worthy of your trust, you have multiple reasons to set your mind against any novel technology that they are putting into the market, or as some activists have put it, "forcing farmers to buy". Yet as Neil de Grasse-Tyson has said, and can be seen on many t-shirts and coffee cups:

> *"The good thing about science is that it's true whether or not you believe in it."*

I put that to Mark and he immediately countered: "Yes, that is quite appealing but nevertheless, absolutely facile". Any new technology requires consumer acceptance and to most, the science is irrelevant. As he pointed out, the early molecular biologists were somewhat circumspect about what may be ultimately possible with their ability to make new genetic combinations. "They set up a regulatory ratchet that eventually strangled them. They couldn't dial back from that."

Yet he told me that after writing a number of pieces specifically against the evils of Monsanto, and all large corporations, he received some correspondence from a reader asking why he was rejecting GM technology based on the simplistic notion that because a large corporation was selling the outcomes, it was not good for the planet. Was he against wheels just because they were sold by large multinational corporations? That led him to question his own motives and beliefs. The realisation that he drove a car, used a computer and telephone, and flew on jet aircraft made by large multinational corporations helped make the realisation that you cannot be against everything for the simple reason that it was made by a corporation, large or small. Even perhaps if that large corporation was the infamous Monsanto.

So, as he had brought up the M-word, I quickly asked him whether he has ever had anything to do with Monsanto:

> *"Yes, as you can imagine it is quite a small pond, and I do run into Monsanto people from time to time. Often it's at a conference or some sort of media event. And occasionally they invite me out to dinner afterwards, which is always fine. Nevertheless, I always make sure I pay for my own meal. You know what perception can be, and I don't want there to be any perception that I am in the pay of Monsanto."*

Of course, that has been the accusation levelled against him on numerous occasions, especially so after his speech to the Oxford Farming Conference. That accusation came predominantly from his former "friends". The people he had sold out. Sold out their values and beliefs.

This is a phenomenon widely experienced by those who espouse the benefits of GM crops and foods. We are all "shills". The first time I was called a shill I had no idea what it meant, but the way it was said I was able to fairly clearly formulate an assumption that it didn't mean handsome, erudite or inspiring. Wikipedia tells me the term arose in 20th century USA, and usually involved some sort of confidence trick or advertising of a particular product, person or organisation without revealing that one had a personal or fiduciary relationship with said product, person or organisation (for which you can mostly read Monsanto in the GM crop space). For the record, I have tried a number of ways to get in contact with Monsanto to ask some questions for this book, and sadly, none of them came to any sort of fruition. In fact, I even asked one of my friends and former students who used to work for Monsanto, whether I could ask her some questions for this book. She told me, "It's not that I don't like you Ian, I'm just circumspect" (anonymous by personal request – now I feel like a real investigative journalist, I'm protecting my sources). Not only that, one day Monsanto may still like some of my research enough to fund me to do some more. For the record, Monsanto have never paid for my dinner either. Although come to think of it, I recall I got a free Monsanto pen when I was at a crop science conference in Minneapolis in 2000. It was one of those hopeless ones that popped apart in the middle about a month later and the spring flew out the window, never to be seen again. Whereas the Pioneer HiBred pen – well that was an absolute cracker and was made of plastic containing biodegradable maize starch. I still have that one. And a matching key ring in the shape of a maize cob. Probably I should declare it somewhere but I haven't. Does that make me a shill? I've got a couple of friends who used to work for Pioneer, so maybe I'll ask them.

When activists don't have a grasp of the facts, the numbers, the skill or the background to attack the ball, they attack the man, who in many cases may be a woman. They don't discriminate. These attacks are an occupational hazard for some of the most pro-GM warriors. Hard-working wonderful scientists and science communicators like Alison Van Eeneenam, Kevin Folta, C. S. Prakash, Wayne Parrott, Peggy Lemaux and Rob Fraley. No, I won't include Rob Fraley, not because he isn't hard-working and all those other things, but he does work for Monsanto (for close to 40 years). So I suppose he can never be called a shill in the strict sense, yet I am sure he has been labelled with that moniker.

These sorts of attacks are called *ad hominum*. Like patriotism being the last resort of a scoundrel, *ad hominum* attacks come into play when you cannot refute an argument with fact or scientific method, so you instead attack the character, circumstances or motives of the opponent. It could be said that, in this space, none know the sensation of enduring such attacks as does Kevin Folta, pro-GM activist, Twitter-God (almost 20 000 followers) and Chair of the Department of Horticultural Sciences at the University of Florida.

Through the US Freedom of Information Act, an organisation called Right To Know (RTK) targeted the e-mails of a number of pro-biotech scientists at various public universities. RTK say they are "Pursuing Truth and Transparency in America's Food System", which really sounds like quite an admirable objective until you read some of their articles, like: "My friend died from cancer today" and "Food Evolution GMO film serves up chemical industry agenda". How? By showing that virus-resistant papayas and bacterial wilt-resistant bananas reduce the need for pesticides? RTK hit gold when they uncovered a $25 000 payment to the University of Florida to assist with travel costs when Kevin was giving his pro-GMO talks to schools, universities, clubs and societies around the USA. Here was a man who was demonstrably a shill! One thing RTK did omit from the argument is that they are funded almost entirely by the Organic Consumers Association, an organisation whose stated aim is to have "a global moratorium on GM crops and foods", which is a little different to pursuing truth and transparency in food production. Then the story was published by *The New York Times*, with the headline in an article written by Eric Lipton, 5 September 2015, and in addition they posted the contents of thousands of e-mails online.

Kevin is a busy man, as you can imagine. We tried on numerous occasions to catch up, and finally got a time and date that suited us both. It was in the morning of 11 September 2017. During the day and night of 10 September, Hurricane Irma crossed the Florida coast and headed up the peninsula. Irma caused havoc across the Caribbean and the south-eastern part of the USA, killing over 146 people, over half of whom were in Florida. A record of 6.5 million Floridians had to evacuate their homes, leaving over $50 billion of damage in the USA alone. Kevin and his university were without power for some time, and we have never caught up since. Kevin being Kevin, he got even busier. He continued a very successful podcast *Talking Biotech*. Just to make his life a little busier and more interesting, he sued *The New York Times* and the author of the article, Eric Lipton, for defamation. Many people are watching what transpires.

In the little town of Frick, on the train line between Basel and Zurich in Switzerland, lies a quite amazing and wonderful agricultural research station. This particular institute is called the Research Institute of Organic Agriculture, for which the acronym in German is FiBL (pronounced feeble). I was lucky enough to visit there in 2017 and meet the charismatic Institute Director, Professor Urs Niggli. For the record, he did buy me dinner, and beer. When I say beer, it was a local version called Feldschlossen. So not really beer, but it's the thought that counts. So if I sound like a shill for organic agriculture, I do solemnly swear it was nothing to do with the beer. He didn't join me in having a beer. His very plausible excuse was that he had broken his ankle, as evidenced by the crutches he was using. This meant he was on blood thinners and couldn't partake of alcoholic beverages.

There is nothing feeble about FiBL, and it can also be said that organic agriculture is none too feeble in Switzerland. As we've already discussed, GM crop research does exist in Switzerland, but it takes place in high-security premises under 24/7 video surveillance. This is not the case with organic research. Organic agriculture is booming in Switzerland, with 15–20% of most fruit and vegetable consumption being organic in some form. Some of the success of organic production in Switzerland is underpinned by solid science, and much of it has been led by Urs Niggli and his team. His background was as a weed scientist, with much of his earlier work looking at understanding weeds in pastures. His study included pastures and meadows, and he began to think of his work as "plant sociology", or how plants interact with each other. He started to manipulate plant populations with herbicides and became involved in assessing herbicides for use in public spaces. He told me (echoing Matin Qaim in Chapter 4) that sometimes herbicide use becomes a necessity because of sloppy agronomy, and there are alternatives to herbicide use in some, but not all, situations.

Pastures in Switzerland are not just important for all those rolling green areas you watch as you fly past in the train, or photograph with snow-capped peaks and little quaint villages, but they sustain Switzerland's most economically important herbivore, the dairy cow. Hard to think of a more important job (unless you're a vegan). Where would the world be without Swiss milk? Swiss cow's milk goes into all those delicious cheeses, of which Switzerland produces more than 400 different varieties, such as the famous Emmental, Gruyere, Appenzeller, Tilsit, Tomme Vaudoise, Vacherin and Raclette. As if that isn't enough, Swiss milk is also an essential ingredient in some of the world's best chocolate including Lindt (yes even the 70% cocoa stuff) and other world-class producers like Cailler, Frey, Tobler, Suchard and Camille Bloch. Swiss beer has not taken the world by storm, but Swiss dairy products have. Some of the cheese you get most frequently in the USA, generically called Swiss, is not Swiss at all, and its timid blandness is a reminder that you cannot buy a cheese called "Swiss" in Switzerland for a very good reason. The Swiss just wouldn't eat it.

Urs became the Director of FiBL in 1990. He became drawn to organic agriculture in the late 1980s because he saw that some of the farmers undertaking the most innovative approaches to weed control and soil fertility were the organic and "bio" farmers. He felt that these were the farmers who were embracing diversity. Conventional agriculture, particularly with the use of herbicides, "served the purposes of an industry not developing diversification but simplifying agricultural systems". In 1990 he was pro-organic because to his mind these were the farmers who embraced diversity, used weeds to attract beneficial insects and were forward-thinking with five, seven or even nine-year rotation plans. He did not see GM crops as a way to increase sustainability, but a means to make farming easier in the short term. Of course, pretty much the only GM crops available were Bt insect-resistant plants or glyphosate-resistant plants – so he was absolutely correct regarding the diversity. When Urs joined FiBL, which had been established in 1973, it was not unlike most of the certified organic agricultural sector – a cottage industry with fewer than 20 staff. He has now built it into a sustainable agriculture research powerhouse, with over 175 staff in Switzerland and, more effectively, nodes in Germany, France, Austria and around 60 staff, plus an umbrella organisation in Brussels to liaise with the EU.

By 1990, the organic farmers had formed associations that started to write their own rules and regulations in stone. Urs felt that they started to become too restrictive in what they allowed. Some farmers were of the opinion that it was all about marketing, but as Urs put it: "Everyone knows that the rules, as written, would keep organic as a niche market." Urs was not driven by an objective to make the world certified organic. He wanted agriculture to become more ecologically sustainable, and to his mind, this meant embracing the best possible modern technologies available if they could deliver good outcomes for the farmer, the environment and the consumer. He believed that as long as precision farming, nanotechnology to reduce food waste, new genetic technologies and new fertiliser technologies were ignored, the industry could never move forward to what he terms Organic 3.0. Organic 3.0 is all about developing the most sustainable systems, and not being limited by the most conservative people within the industry, such as the biodynamic farmers. In his own words, "this very restrictive method of farming is not the best one for sustainability". He quietly started a crusade to see what traction he could make with some gentle nudging on new technologies. He felt that the organic industries use fear to get people to support them, and he knows he will never change the opinions of the "holy ones". In 2013, he dared to present some of these ideas at a conference, and made a low-key mention that perhaps GMO technologies should be looked at moving forward. However, he sees his goal as not just changing the organic industry, but producing research outcomes that can be embraced by all sectors of the industry, including conventional farmers. After all, it is not just the black-and-white organic *versus* conventional, there is a huge gradation of farming practices on the spectrum in between.

Then, in 2016, Urs spoke at a workshop for young journalists in Munich. He presented a few ideas for the future of sustainable agriculture. He was met by a young journalist with the words "Urs, I hear you were positive about genome editing", and the very next day, there was an article in the *Tages Zeitung*, an alternative newspaper based in Berlin. The article was written by Jost Maurin, and was entitled (in German) "CRISPR has great potential". This was followed soon after by "Organic scientist says GM crops are the future".

Let's first take a step back. It's not a total surprise that the staff at FiBL have some awareness of new genetic technologies, which will be further explained in the next chapter. A major programme at FiBL is breeding for the organic and sustainable agriculture sector. FiBL has breeding programmes aimed at improving yield and insect resistance in organic cotton, improving cereal quality under low soil fertility, and developing new disease-resistant lupins. Unlike many in the organic industries, FiBL is embracing genetic technologies such as DNA markers and genomics to more effectively harness genetic diversity. Urs and his group do not see themselves as merely serving organic and biodynamic niche farmers.

> *"We need to make progress in the conventional sectors as well. Less input – more biology. How can we help crops to defend themselves, and uptake nutrients more effectively. If we don't have a debate, and try to be more sensible and rational, we won't be able to stop the rise of general hysteria against the new breeding technologies."*

Shortly after the "Organic scientist says GM crops are the future" headline, his phone started to ring incessantly. Initially, Urs was amused by the whole situation, but then "the shit storm went national and international". Journalists and activists were ringing him all day and night. His co-authors and supporters within FiBL and other European networks retired from the conversation. He then started to receive indignant calls from certified organic and biodynamic authorities, some of which threatened to withdraw research funding to FiBL. It became so stressful that in March 2017, he decided to not give any more interviews on the matter. While this worked, journalists then started to ignore him. Previously he was the "go-to person" for issues regarding soil health, sustainability and the effect of climate change on agriculture, but he felt he had fallen off the list. When I met him in Frick, he was preparing to go to a meeting in Darmstadt to show cause to FiBL's biodynamic donors as to why they should continue to fund him. While biodynamic farming is the most restrictive in its practices, Urs feels that while they will never share his views, they are tolerant of them.

Overall, Urs is fed up with the unscientific arguments of the anti-GMO NGOs and some of the organic organisations. He sees a wavering of hostility in some organisations, some of whom are willing to sit down and discuss the new technologies available. Bioland, Germany's largest organic certification scheme, has had discussions with scientists at the Max Planck Institute in Berlin regarding new breeding technologies (NBTs). Having been in the room with activists many times at these debates, Urs has come to the realisation that when he argues that GMO and NBTs could be integrated with sustainable farming on a case-by-case basis, a certain look comes over the faces of the activists. It is the look of fear. While what he is saying is definitely challenging their values, there is something else he is challenging more. He is challenging their very livelihoods. Many of the NGOs need to continually portray GM crops as intrinsically evil, a tool of major corporations and environmental vandalism. They need to extend that fear to the products of NBTs. This has become a fight for survival for many of these organisations. Yet as Mark Twain said:

> *"Loyalty to petrified opinion never yet broke a chain nor freed a human soul in this world – and never will."*

REFERENCES

1. B. J. Haas, *et al.*, *Nature*, 2009, **461**, 393.
2. E. M. Goss, *et al.*, *Proc. Natl. Acad. Sci. U. S. A.*, 2014, **111**, 8791–8796.
3. M. Lynch, *Scotland, a New History*, Pimlico, London, 1991.
4. G. L. Carefoot and E. R. Sprott, *Famine on the Wind*, Rand McNally, Chicago, 1967.
5. A. Drenth, I. C. Q. Tas and F. Govers, *Eur. J. Plant Pathol.*, 1994, **100**, 97.

6. S. B. Goodwin and A. Drenth, Origin of the A2 Mating Type of Phytophthora infestans Outside Mexico, *Phytopathology*, 1997, **87**(10), 992–999.
7. A. Koeppel, E. B. Perry, J. Sikorski, D. Krizanc, A. Warner, D. M. Ward, A. P. Rooney, E. Brambilla, N. Connor, R. M. Ratcliff, E. Nevo and F. M. Cohan, Identifying the fundamental units of bacterial diversity: A paradigm shift to incorporate ecology into bacterial systematics, *Proc. Natl. Acad. Sci. U. S. A.*, 2008, **105**(7), 2504–2509.
8. G.-E. Séralini and J. Douzelet, The Taste of Pesticides in Wines, *Food Nutr. J.*, 2017FDNJ-161.

New Kid in Town

*"Nothing in life is to be feared, it is to be understood.
Now is the time to understand more, so that we fear less."*
Marie Curie

"Nothing has changed; everything is new."
Sjömagasinet motto

Sjömagasinet is a restaurant in Gothenburg, on Sweden's lovely
and rocky west coast. The restaurant is owned by Ulf Wagner,
and with Sweden's Chef of the Year 2010 Gustav Trägårdh, they
have created a restaurant in a big old red shed "inspired by the
fruits of the sea". Literally, Sjömagasinet translates to something
along the lines of Sea Magazine. According to the 2018 Michelin
Guide, Sjömagasinet is:

*"A charming split-level restaurant in an old East India
Company warehouse dating from 1775; ask for a table on
the upper floor to take in the lovely harbour view. Cooking
offers a pleasing mix of classic and modern dishes; lunch
sees a concise version of the à la carte and a 3 course set
menu."*

Good Enough to Eat? Next Generation GM Crops
By Ian D. Godwin
© Ian D. Godwin 2019
Published by the Royal Society of Chemistry, www.rsc.org

Given there are no structures in Australia from as far back as 1775, the building of itself created a wonderful historic feel for me. Naturally, for one who frequents the West Toowong Bowls Club for a meal out, being at a restaurant such as this was a treat in itself. In summer. With seafood. In Gothenburg with its lovely harbour, beautiful old cobble-paved centre and craft beers (with ships on the labels).

The whole experience was made a Double Treat because I was there with a bunch of fellow plant scientists for a conference dinner, led by the irrepressible and charismatic Stefan Jannson. Stefan is a Professor and the Head of the Department of Plant Physiology at Umeå University in northern Sweden. Umeå is about as close to the Arctic Circle as universities get, without actually being inside it, like Trømso, Murmansk and Reykjavik are. Stefan was on the conference organising committee for Sowing the Seed: New Breeding Technologies in the Plant Sciences, a meeting set up by the Society for Experimental Biology in July 2017. For a Triple Treat, he had grown some CRISPR-edited cabbage in his garden in Umeå, transported it to Gothenburg, and talked the award-winning chef at Sjömagasinet to produce a main course of "Poached cod with a terrine and foamy bullion of suckling pig CRISPR cabbage and a potato purée with deep fried onion and thyme" (see Figure 10.1).

We were not the first people in the world to eat CRISPR-edited food. That may have been Stefan when he grew and cooked some for a radio science show in Sweden in 2016. Another claim of precedence on that matter was Yinong Yang's lab at Penn State University, who had a bit of a party with their CRISPR-edited non-browning mushroom, and then there is always Alison van Eenennaam's cattle genetics group at University of California, Davis who had a lab barbecue in late 2016 courtesy of one of their CRISPR-edited hornless cattle. If we put the three together, all we would need would be a CRISPR-edited potato and it has the makings of a good meal. And just down the road at SLU Alnarp they had CRISPR-edited potatoes, although these were aimed at the industrial starch market, so perhaps not ideal for a meal at a restaurant.

However, I am confident that this wonderful meal, eaten upstairs in the lovely wooden-beamed warehouse now known as

Figure 10.1 The menu for the first ever CRISPR gene edited meal served in a Michelin-starred restaurant – Sjomagasinet, in Gothenburg on the west coast of Sweden.
Autographed by Professor Stefan Jansson, Umea University plant scientist and purveyor of the finest CRISPR-edited cabbage.

Sjömagasinet, was the first meal with a CRISPR-edited ingredient cooked to perfection and served in a restaurant (with matching wine). It must be true. Stefan said so.

CRISPR editing? Yes, it does rather cry out for a spot of elucidation, doesn't it? It stands for Custom Removed or Inserted Sequences of Preferred RNA. Or actually it could be an acronym for Create Really Intelligent Special Proteins Readily. Or Consistently Reliable Interruption or Significantly Promoted Regulation of gene expression would also be a good one. Alas, 'tis' none of the above. It's actually Clustered Regularly Interspaced Short Palindromic Repeats. Brilliant in its conception really. The clarity of what CRISPR is or does will always be obfuscated by the acronym. When I think of Clustered Regularly Interspaced Short Palindromic Repeats, it conjures up words like Eenennaam, which has lots of regularly interspaced short repeats, and like the CRISPR acronym is almost impossible to remember the spelling. If I recall correctly, Alison van Eenennaam told me over beers at a conference in San Diego that it means something like "any name", and was proffered as a surname to invading Spanish/ French or something like that when they were trying to hold a census (perhaps in mind of collecting taxes in future). OK it was over beers. And jetlag. Anyway, whatever it means it's a lot better name than Clustered Regularly Interspaced Short Palindromic Repeats, but we are stuck with that now. So what does CRISPR really mean, and why should you care?

Well, you should care because this technology is going to drive ... actually, this technology *is* driving a biological revolution. *Agrobacterium* drove a plant biological revolution. CRISPR is driving a biological revolution. It's life Jim, but not as we know it. This is a biological phenomenon (read biotechnology) that allows us to alter the way in which genes are expressed. We can do it in petunias, irises, proteas, Scots Pines, crustaceans, mammals (that's you – yes you, and wombats, cattle, sheep, bears, camels, dugongs, gazelles, mongooses and mongeese and gorillas and guerrillas), insects, bacteria, archaea, yeasts, fungi, and pretty much every organism on the planet. CRISPR can cure any known genetic disorder in any organism. Even the ones (organisms and disorders) that are yet to be discovered. CRISPR can create any known genetic disorder in any known organism. That's scary! But look at it this way – in domesticated plants and animals, this is what we do. It's what we call domestication. Imagine! But I'm getting ahead of ourselves. Or you're getting ahead of myself.

The CRISPR system is based on what could be termed a natural bacterial immune system. Now, have a bit of a think. Your immune system is all about defending yourself against the major pathogens – bacteria and viruses. If you happen to be a bacterium (and well done for learning how to read anything other than a DNA sequence), it's exactly the same. Your main enemies are ... bacteria and viruses, which are biotic stresses. Well ... there's also smoking and alcohol and heat and freezing and salt and rotenone and copper sulphate, but these are abiotic stresses.

The CRISPR system was independently recognised by a number of groups in Japan, the Netherlands and Spain, where various aspects of the bacterial immune system were discovered. The major feature of the systems identified in different bacteria were the presence of interrupted/interspersed repeats of DNA sequences. These appeared to be some sort of recognition of the DNA/RNA of the nasty little microorganisms that tended to attack other microorganisms. Then other groups looked further into these potential immune systems. What they found were a family of CRISPR-associated proteins – actually enzymes that seemed to share one quality. The CRISPR-associated enzymes (remember enzymes are proteins), seemed to share the ability to cut DNA. Hence the system became known as CRISPR/Cas, meaning Clustered Regularly Interspaced Short Palindromic Repeats/CRISPR-associated proteins. The most studied version of the system was in *Salmonella pyogenes*, which is associated with a protein called Cas9. It's called Cas because it's a CRISPR-**as**sociated protein. The gold standard system at the time of writing is now known as CRISPR/Cas9, and it has been used to edit a huge number of genes. Let's spend a little bit of time to understand how the system works, and what its outcomes may be.

To understand what CRISPR does, you first need to envisage that the system has a means of identifying a specific DNA sequence in the genome. Once it has identified this sequence, the Cas protein (which is actually an enzyme) will bind to the DNA at that site and cut both strands of the DNA (for mechanism see Figure 10.2). This is called a lesion, and like all lesions in DNA, the host has the means to repair the lesion with its own DNA repair enzymes. Basically, the types of genomic

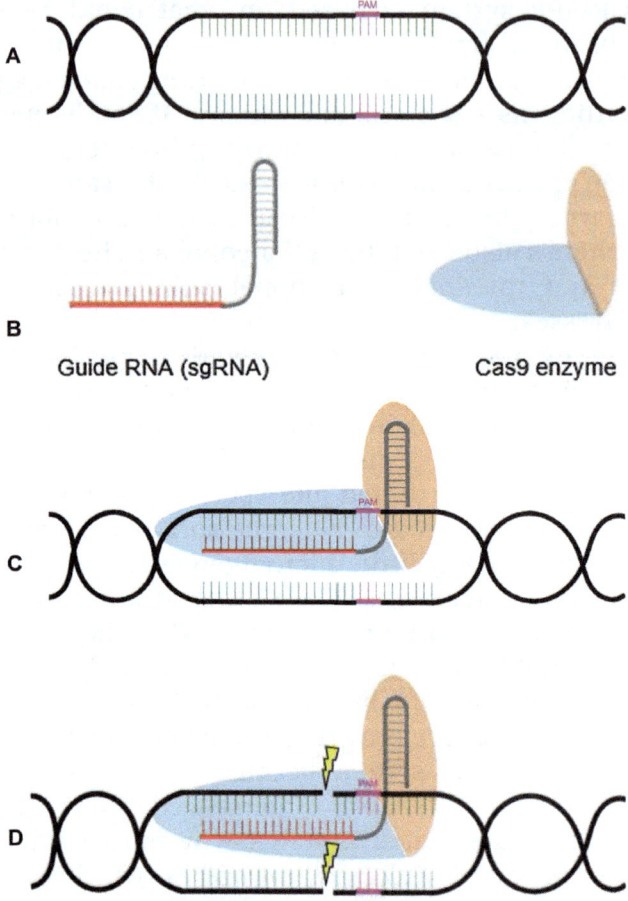

Figure 10.2   Cartoon of how CRISPR makes a lesion in DNA. (A) The genome target site is known as the PAM. (B) Two components for the CRISPR system are needed: the Cas enzyme, in this case Cas9, and the guide RNA with targets the PAM site *via* the sgRNA. (C) The sgRNA binds to the target site of the host genome and the Cas9 enzyme binds to the DNA–RNA complex. (D) The Cas9 enzyme makes a double-stranded cut in the DNA.
Photo by K. Massel.

changes that CRISPR can make are divided into two major classes:

1. Non-Homologous End Joining (NHEJ)
2. Homology Dependent Repair (HDR)

NHEJ is the easiest and most commonly used form of CRISPR editing, because once the break is made, the system relies on the natural DNA repair mechanisms, as would happen in any cell.

It may shock you to know this, but DNA repair mechanisms are going like the absolute clappers all the time, repairing all those mistakes that your cells made every time they divide. There are also times when the repair mechanisms have to work really hard, for example when you expose yourself to that wonderful carcinogen, ultraviolet or UV light. It doesn't matter whether you expose yourself to sunlight, or an artificial UV light like a tanning bed (where you are trying to get exposed) or a lab UV source (where you are presumably trying to avoid exposure). In addition to making you sunburnt (or tanned depending on how you wish to view it), that light will cause lesions in the DNA of your skin cells. The DNA repair mechanisms will be working overtime to correct these lesions. Sometimes they are so busy they will make mistakes, much as humans do when texting and driving at the same time, or sometimes just when they are only texting or only driving. What happens when DNA repair mechanisms make a mistake? Well they will sometimes put the wrong nucleotide into a DNA strand, substituting a T for a C, for example. Other common mistakes include actually creating small deletions, like just totally missing a single nucleotide, or insertions, such as adding a single nucleotide. The outcomes of these mistakes could vary from actually changing nothing, through subtle small changes, to totally changing the action of the gene. The most disruptive change that can be made is knocking out the expression of the gene altogether, which is not an uncommon outcome when a stop codon is inserted into the amino acid sequence and no active protein is made.

More complex, harder to achieve, yet giving you infinitely more new genetic possibilities is the HDR version of editing. The difference here is that you provide a template of the DNA sequence you would like to insert into the lesion while it is repaired. The sequence you provide could be a single nucleotide or could even be hundreds of nucleotides long. The aim is to insert that into a gene sequence to significantly change the amino acid sequence it encodes. In practice, while achievable, it is achieved at much lower frequency because you are now relying on perfect DNA repair with your sequence incorporated perfectly.

There are other systems that have been used to make these types of genome edits, most notably zinc finger nucleases (ZFNs) and transcription activator-like effector nucleases (TALENs). These have been used to make genome edits, yet they are significantly harder to use, and not always as reliable as the CRISPR/Cas9 system. Together, these can be lumped together as genome editing techniques, or, if you are a plant or animal breeder, the term new breeding technologies is often bandied about. New breeding technologies don't only include genome editing, however, I suspect it is often used to describe genome editing from a marketing stance.

Let's consider a couple of examples of how these techniques have been applied, a horny one and a fragrant one.

As we all know, cows have horns. If they didn't, biodynamic farmers would have to change the formulae of many of their magic potions. Alison van Eenennaam from UC Davis told me about their genome edited cattle (at least one of which they ate at a lab BBQ). The animal was hornless, or what is called polled. Polled is a desirable trait in domesticated cattle for some obvious reasons (Figure 10.3). Firstly, I think it should be fairly obvious that polled cattle are much safer to work with in the farming situation. Secondly, polled cattle are much less likely to damage one another. This damage may be on purpose, because the cattle are having a little disagreement over a prize patch of clover or a good-looking heifer. However, it is just as common that it is accidental, and brought about when herding and concentrating cattle into a yard (for things such as veterinary treatments and

Figure 10.3 Holstein Friesians, polled cattle, usually by manual dehorning.
© Elisabeth Aardema/Shutterstock.

milking) or loading onto a truck to take to another farm, the local agricultural show, or even to market.

Companies such as Wal-Mart, Starbucks and Nestlé encourage the use of polled cattle from an animal welfare perspective, and will pay premiums for dairy products sourced from polled herds. So there are quite a few incentives to have a polled herd of cattle, and many livestock owners have been breeding for the polled trait for many generations. The major problem has been that the polled trait, encoded by a gene called POLLED, is dominant. As a result, when you select or buy polled cattle, those cattle may be heterozygous for the trait. Another way to put it is that they are "carriers" of the horned allele, and hence if you mate a polled bull to a polled cow, some of the progenies (about a quarter in fact) are going to have horns (Figure 10.4). There is another issue to consider here, particularly in dairy herds.

Most dairy herds around the world are the familiar black and white Holstein Friesians (Figure 10.3). For some reason these are known as Holsteins in North America but are most commonly referred to as Friesians in Australia and the UK. The POLLED gene is at very low frequency in Holstein cattle, and for good reason. The gene is linked to genes that have a deleterious effect on milk productivity. For every lactation cycle (calving event which leads to milk production), the polled females will yield around $250 less milk than the horned females. This is known as

Figure 10.4 The Texas Longhorn, an example of non-polled cattle.
© Kristin Shaffer/Shutterstock.

"linkage drag", a well-known problem for plant and animal breeders, when a "good" gene is closely linked to a "bad" gene that has some sort of deleterious effect on productivity or resilience. Hence the polled gene itself does not lead to a reduction in milk production, but the genes that come with it on the same chromosome do. Here is an obvious target for genome editing.

A small company known as Recombinetics from Minnesota produced the first genome-edited polled cattle in 2015.[1] They edited the Holstein gene by inserting a version of the gene known as POLLED$_C$, which had a 212 bp insertion in place of a 10 bp deletion. The POLLED$_C$ version of the gene was from Celtic cattle races. They produced two healthy calves, known as Spotigy and Buri, with no horns. This demonstrated that it is possible to use gene editing technology to alter traits in cattle without introducing other deleterious traits.

Now for the fragrant version of a gene. Those of you who are rice aficionados will have particular favourites that you like to eat. For many people, especially if you're a fan of Asian foods, these are often the fragrant rices, known more commonly as jasmine rice and basmati rice. There are some quite striking differences between the fragrances of these rices, and the fact that their relative attraction is in the nose of the beholder.

Thais are very proud of their jasmine rices, but ask a Thai about basmati rice and many will tell you that it smells like a dead mouse or rotten hay. Whereas give a Pakistani or Iranian basmati rice and they will rave about the fine, nutty and sweet fragrance. Assail a Pakistani with a jasmine rice and they may proffer the feedback that it smells like perfume or chemicals, and it really put them off the taste of their Sindh biryani. In fact, a 2010 study using a trained sensory panel at the International Rice Research Institute (IRRI) in the Philippines revealed what real rice experts think about rice from different geographical origins. Now when it comes to doing taste testing, you want the best, right? Well this study was led by a scientist called Elaine Champagne, so it doesn't get any better than that does it. Her co-workers included the absolutely stellar Claude Lobster, Clarice Caviar and Jean-Baptiste Fois Gras. Well, it may surprise you to know that I just made those names up, and if your name is actually Claude Lobster, dear reader, then please accept my apologies. In actual fact, some of the other scientists involved were my University of Queensland colleague, rice food scientist, Melissa Fitzgerald and a former collaborator, Russell Reinke, who used to run the Australian rice breeding programme before he went off to IRRI. Melissa was at IRRI before she returned to Australia to take up a professorship in Queensland. Elaine T Champagne is actually a real person, and is a food scientist working with the USDA.

But back to the Champagne *et al.* (2010) study about rice.[2] Overwhelmingly, the sensory panel said the aroma they most associated with the Thai jasmine rice was "popcorn", which is an interesting phenomenon. What makes it more interesting is that a similar aroma is also sometimes associated with lobster (bravo Claude), and other foods too, like the Asian spice, pandan (from pandanus leaves), and fried prawns. Meanwhile the same panel found the Pakistani basmati rices to have more "hay-like", "musty" or "grassy-green bean" aromas. However, the main aroma compound of both the jasmine and basmati rices is in fact a volatile compound known as 2-acetyl-1-pyrroline. This particular compound was not present in the rices from other geographical origins. People obviously like the aroma, because they are willing to pay more for it. Most Australian rice is marketed by a company called SunRice. They import and package up jasmine and basmati rices as well.

In the interests of full disclosure, it does also rest upon my shoulders to reveal that the rices of Australian origin seemed to unite the tasting panel, but not always in a good way. One flavour/aroma characteristic of these rice varieties was "sewer" or "animal". Drilling down into the data, a number of panel members also used the term "piggy" to describe the Australian varieties, which may actually be more insulting to pigs than it is to Australian rice breeding and/or tastes. Melissa Fitzgerald assured me that the overwhelming characteristic of the Australian rices was that they were "sweet tasting". So perhaps the aromas were more like that of a marzipan piggy? I now seek to sweep this finding under a mat somewhere, yet it does rather beg the question: Russell, how did you feel putting your name on this paper, given that these rices presumably came from your breeding programme?

2-acetyl-1-pyrroline. Sounds like a chemical, doesn't it? Well that's because it is, just like some of the other chemicals in rice like amylose, amylopectin, phytic acid, linoleic and linolenic acids, as well as palmitic acid, and a bunch of different types of prolamins, glutelins and albumins, which are all proteins. Rices also often have other volatile compounds contributing to their aroma, like the delicious and nutritious (E,E)-2,4-decadienyl, as well as hexanal, nonanal, 4-vinyl-guaiacol and 4-vinylphenol. Who would ever be so stupid as to consider eating such a combination of chemicals? Well, somewhere between three and four billion people do exactly that every day and sometimes more than once. My rice breeding colleagues, and my friends at SunRice, I have a cunning plan. It's along the lines of the Non-GMO Project. The Non-GMO Project? It's the label that promises "Everyone Deserves an Informed Choice". Then they go out of their way to misinform you. It's like those labels that say "hormone-free beef" when nobody has used hormones in beef for the last 30 years, if ever. Or those "organic, chemical-free" labels, like on the tube of zinc cream at the hippie surf shop. Meaningless drivel, and all about misinformation to create fear, and develop a marketing edge. Marketing edge = we sell more and get a higher price, even though our product is no different to the one without the "bullshit-free" label. When it comes to selling certified non-GMO salt or mineral water, you know it's all about marketing misinformation.

So, I'm going to offer this to you Australian rice marketers for free. Let's make some stickers to put on your non-fragrant rice that proclaims: "Our rice is proudly 100% 2-ACETYL-1-PYRRO-LINE-FREE". Those fancy-schmancy, hoity-toity, jasmine rice people won't know what hit them, and they won't be able to use the label or we will sue them! With a clever marketing campaign this one is going to go gangbusters. We'll get the *Huffington Post*, *The Guardian* and maybe even *The Betoota Advocate* to write up some food and health articles and watch the sales of non-fragrant rices absolutely boom. Suddenly, everyone will be "afeared" of the dreaded 2-acetyl-1-pyrroline, and it will take years for scientists and nutritionists to convince people to remain calm. Meanwhile hipster vegan tapas bars will have blackboards outside proclaiming that their soy-cheese arancini and their biodynamic kale and kaffir lime risotto are "100% 2-acetyl-1-pyrroline free". They can even make the claim that their turmeric lattes and quinoa Bircher muesli are also "100%

2-acetyl-1-pyrroline free". The claim can even be extended to the pink salt, you know that stuff from the Himalayas that is 250 million years old but has a use-by date of January 2020.

OK, back to the real story. Where were we? Oh yes, 2-acetyl-1-pyrroline of course. My mate, Robert Henry, the Director of the Queensland Alliance for Agriculture and Food Innovation at UQ, was the first person to tell me about 2-acetyl-1-pyrroline, over a beer. In fact, Robert is the nerd's nerd. He taught me all sorts of stuff about food. Well, mostly beer actually. Chill haze (tick). Flavour components from hop trichomes (tick). Limit dextrinase (tick). Surfactants in beer (tick). Surfactants? Yes. Otherwise you wouldn't get the froth on top of your frothy. Relax, it's a natural component of barley. Or was it hops? In fact, he's told me so many geeky things about the chemistry of food and beer while imbibing beer or red wine, but for some reason I, for the life of me, can't remember most of them. For a man who started his career as a cereal guy, he has expanded to work on pretty much every plant you can imagine. When not being a plant geek, he's either sailing or being a plant geek – like hunting for new species of rice in crocodile-infested waters in Northern Australia. Robert's lab (when he was at Southern Cross University in the subtropical epicentre of the Australian marijuana industry) identified and cloned the gene responsible for the production of 2-acetyl-1-pyrroline in fragrant rices. The work was performed mostly by two young PhD students, Louis Bradbury (now at New York University) and Tim Fitzgerald (now a patent attorney), and they found that the fragrance gene was actually a bad gene. It was a bad gene in two ways. Firstly, it was a recessive version of the gene, or what we call a null allele, because it had undergone a mutation such that it no longer encoded a protein. Secondly, what the gene was supposed to encode (when it was a good gene) was an enzyme known as betaine aldehyde dehydrogenase, hence the gene was known as BAD2.[3] So in summary, if you were breeding for fragrant rice, all you needed was a bad version of BAD2 and everything was good. Well, fragrant at least.

Caixia Gao is a scientist at the Chinese Academy of Sciences in Beijing. Some of my students met her at a genome editing meeting on the south coast of New South Wales in 2017, and they were blown away by the research she was doing. In fact, she is still known as the "crazy CRISPR lady" in my lab. Crazy in the nicest possible innovative way for somebody leading the charge with disruptive genetic technologies. I spoke to Caixia in early 2018. She told me that when she finished school, like everyone else, she wanted to do something important, like medicine or engineering (like both her parents). However, she didn't get the right scores and "wasn't the right personality" to get into prestigious schools like Peking University. Wasn't the right personality? . . . .

So she ended up at an agricultural university, and once there found out she had a flair and passion for plant science research. After completing her PhD, she got a job at Denmark's largest seed company, DLF Trifolium, who work predominantly on pasture and forage species. Caixia worked on genetic transformation of pasture grasses such as *Festuca* and was very successful. However in 2010, when she was offered the opportunity to move back to China to lead her own lab at the prestigious Chinese Academy of Sciences in Beijing, she jumped at the chance. She joined the Institute of Genetics and Developmental Biology and was charged with leading wheat transformation. In a few short years she had moved from transformation to genome editing with ZNFs, then TALENs. Finally, when Jennifer Doudna's group from Berkeley published on the use of CRISPR/Cas9 technology in 2012, they decided to try it, although they had serious reservations about whether this technology would work in plants. Within a year, papers started to appear in journals, and so for a while, the Gao lab worked on both TALENs and CRISPR technologies. They published their first wheat editing paper in 2014 using TALENs, but within a couple of years, everyone in the lab was really working on CRISPR. It was simpler, more reliable, and above all, much quicker to generate edits of targeted genes. Now in her lab of 16 staff and students, she has successfully achieved genome editing in *Arabidopsis*, wheat, tomato, lettuce, maize, rice, tobacco and potato. Much of what she does is performed as a service for plant breeding programmes and plant science researchers. Given that the easiest edits to achieve were gene knockouts, or to put it another way, turning a gene that encodes a protein into a null allele which can no longer encode a protein, it wasn't long before the rice breeders came knocking on the door. They asked if she could make a fragrant version of one of their elite rice lines, by knocking out the BAD2 gene. Within a year, the Gao lab produced a fragrant version of this elite rice, and the breeders liked it so much they asked her to make fragrant versions of many more of their favourite rice varieties. And naturally, the "CRISPR lady" has delivered.

As mentioned above, it was the seminal paper from 2012 in *Science*, where Jennifer Doudna from University of California (Berkeley), and Emmanuel Charpentier, then at Umeå University in Sweden, published the first report of successful DNA editing with CRISPR/Cas9. Doudna and Charpentier had met at a conference in Puerto Rico, and decided to work together on an interesting enzyme known as Cas9. By 2012, they were able to demonstrate that they could cut DNA at any site in any gene they wanted with this enzyme. They had achieved this using bacterial (prokaryotic) DNA and wanted to demonstrate that the same thing could be achieved in eukaryotic systems, most crucially in humans. The University of California had filed their patent for gene editing using CRISPR/Cas9 in May 2012, publishing in *Science* the very next month. Emmanuelle Charpentier, originally from France, was at Umeå University from 2009 to 2014, and in 2015 she and Doudna shared the prestigious (and lucrative) $3 million Breakthrough Prize in Life Sciences for their CRISPR discoveries. Charpentier was an example of a modern scientist, one who moved around to work on what interested her and get the funding and infrastructure necessary to do so. Having started her career in Paris at the Pierre and Marie Curie University, then the Pasteur Institute, she moved to the USA at three different institutions, then to the University of Vienna, and was at Umeå by the time she started to work with Jennifer Doudna. In 2015, Charpentier took up a position as the Director of the Max Planck Institute for Infection Biology, based at the prestigious and historic Charité Hospital in Berlin.

Meanwhile over on the east coast, Feng Zhang's team from the Broad Institute, a medical research centre jointly involving Harvard and the Massachusetts Institute of Technology (MIT) in Cambridge, Massachusetts, published in January 2013 that they had successfully achieved human genome editing in human cell cultures. The publication came out shortly after filing their patent in December 2012. Hence a casual observer would look at that and probably formulate an opinion that having filed their patent seven months after the Doudna/Charpentier team, they did not have "priority", this being the act of getting in first. However, the Broad Institute team paid a fee in the USA to fast-track their application on the use of CRISPR/Cas9 in eukaryotic cells, and their patent was issued in April 2014, before the UC Berkeley patent. And that's when things started to get a little ugly.

At present, the dispute is ongoing. The latest developments occurred in February 2017, when the US Patent and Trademark Office (USPTO) ruled that the Broad Institute patent had priority in eukaryotic organisms. They accepted the Broad Institute's claim that the original UC patent did not give sufficient detail to demonstrate that the CRISPR system could work in eukaryotes. It should not surprise you that this is where the money is to be made, given that all important agricultural species, from unicellular yeast through the crop plants and domesticated animals, are eukaryotes. Further to that, you are using your eukaryotic systems to read and synthesise this paragraph. It is quite possible that the most lucrative markets for the use of CRISPR/Cas9 technology will be medical, meaning it will have its major future applications in humans.

Naturally, the Doudna/Charpentier team did not take this lying down, and the University of California (Berkeley) appealed the decision. Hold on to your collective hats people. Jennifer Doudna was quoted in *Science*, shortly after the original USPTO decision, as saying: "They have a patent on green tennis balls. We will have a patent on all tennis balls. I don't think it really makes sense".[4] In July 2017, the California team filed an appeal stating "its patent was broad enough to cover all uses of the technology". In their appeal, they also pointed out that six other labs had used their 2012 *Science* paper to achieve eukaryotic editing before the Broad Institute had filed their patents.

In January 2018, the Europeans fired their first salvo in the dispute, and it was not merely across the bows of the Broad Institute, it was a direct hit. If you're a fan of cycling like me, you may hear the acronym EPO and think "cheating". EPO (erythropoietin) was used (illegally) by many cyclists to enhance their performance and endurance, particularly in the long duration "Grand Tours". The drug has the effect of increasing the red blood cell count, and hence allowing for more efficient exchange of oxygen and CO_2. One of the side effects experienced by cyclists is that the red blood cell count made their blood so thick they sometimes experienced circulatory problems. On occasion, this became lethal and a number of cyclists passed away while asleep as their pulse rate slowed down and actually stopped. The only Danish cyclist to ever win the Tour de France, Bjarne Riis, the yellow jersey champion in 1996, went on record to say he had used the drug for a number of years. When asked about side effects he said "the only effect was that I was riding faster".

There is another EPO, and that is the European Patent Office. In January 2018, the EPO revoked the patent granted to the Broad Institute, as their claims did not have priority over the UC Berkeley patent. BOOM! However, the decision did not automatically mean that the patent rights rested with UC Berkeley either. In Europe there are many situations where licences are required from multiple partners for the enabling technology to make a product. Between April and September, 2018, the Federal Appeals Court in the USA considered the appeal made by UC Berkeley. On 10 September, they dismissed the appeal, in favour of the Broad Institute who made the first edits in mammalian cells. This may be the final chapter in the great CRISPR patent debate, but don't count on it. In the meantime, researchers like me all over the world are scrambling to find out whether our licence agreement to use CRISPR/Cas9 in our research will create problems if we got our licence from the eventual "loser" of this mega-patent battle. Further to that, all over the world there are investors in companies such as Editas (founded by Zhang and the Broad Institute) and Intellia (founded by Doudna) who have some serious money tied up into the future of the CRISPR patents and their licensing. In the agriculture field, the main interest has come from the existing biotech companies, who are entering into licensing agreements with one or both of the plaintiffs, to be sure to be sure.

When it comes to new kids in town, CRISPR (and other gene editing technologies) is certainly a revolution in biology, and will soon become part of medical and agricultural advances. However, in the past decade, another new kid has emerged alongside the gene editing technology. That new kid is a thing called synthetic biology. What is synthetic biology? Well it must be important because *Nature Communications* even has a synthetic biology editor, and his name is Ross Cloney. I first met Ross in

Copenhagen, where he was visiting for a week to take a workshop on scientific writing for young researchers at the Danish Technical University. One of the first things anyone notices meeting Ross for the first time are his tattoos. His tattoos proclaim megageek loudly and clearly – I mean, a DNA sequence snaking up his forearm? Ross has separated his tattooed arms into a binary divide. On his "mechanical" arm, he has a depiction of gears that wind and unwind DNA constantly. His "philosophical" arm includes words from a poem: "It is your responsibility as a good person to improve the world".

Ross is based in London, and has a background as a molecular biologist. For most of his research career at the University of Sussex, Ross worked on understanding the mechanisms of DNA repair. DNA repair is, as you will recall, that nearly perfect mechanism for repairing things when something goes wrong. It's only nearly perfect, and hence when it does make a mistake, things like mutations and cancer can be the result. As you have read earlier in this chapter, we also rely on the imperfect nature of DNA repair to get gene edits after we have cut the DNA with a genome editing technique like CRISPR/Cas9. Whilst in the midst of an existential crisis regarding that most of his life as a research scientist had spiralled into such a narrow field that he had become an expert in a phenomenon revolving around three specific amino acids. Is this all there was to life as a scientist? His doubts were brought to a head by the fact that the funding he was working on was also coming to an end. Exactly the sort of things that help many early and mid-career researchers to re-evaluate their next steps. However, rather than ending up as an Uber driver or barista, Ross answered an advertisement to become an editor for *Nature Communications*, and since 2014, that is what he has been doing. It appealed to his broader interests in understanding biology, and put him into the field of synthetic biology, which he describes as "applied molecular biology". Or putting it another way, "taking the parts and processes that have been discovered by basic biology and applying them to achieve desired outcomes". He mainly oversees the manuscripts submitted to the journal on synthetic biology and genome engineering. He firmly believes that these two new kids in town are main drivers of Industrial Revolution 4.0. As he puts it, large industrial-level labs are putting together robotic molecular foundries, which can automate tedious processes such as sequencing genomes, synthesising new genes, cloning and making new constructs. These can all be programmed *in silico*, so that scientists and students do not spend most of their lives on the tedious, but can delve straight into the innovative. I'm not sure whether I share his enthusiasm, because it does rather sound that the biologist of the future is going to become a computer jockey writing R scripts and analysing huge data sets, which makes the human gene project look like a kindergarten.

Wikipedia says that synthetic biology is an interdisciplinary branch combining biology and engineering, which to a degree does mean that it would include genetic engineering. However, what really encompasses synthetic biology is incredibly broad, and the definitions tend to differ widely depending on whether you ask a biologist (like me) or an engineer (and why would you do that?). One of the more commonly used definitions is: "Designing and constructing biological modules, biological systems and biological machines or, redesign of existing biological systems for useful purposes".[5] One of the key technologies underpinning synthetic biology is the ability to make synthetic DNA sequences and then enable their expression to produce synthetic enzymes and other proteins that can be used to develop novel or more efficient outcomes.

Some of the easiest examples to illustrate synthetic biology is our ability to change enzymes such that they become more useful for us. The ability to make an enzyme that has peak activity at room temperature rather than at body temperature many enable us to develop processes to make biological molecules more efficiently. In that same vein of thinking, the ability to also change the pH at which the enzyme functions optimally could also improve its use (and market share if you want to make it for profit). Other options include the repurposing of an understood process to a different outcome, or perhaps the same outcome using a different and cheaper organism.

Take for example, the vanilla ice cream of all desserts – vanilla ice cream. Who doesn't like vanilla ice cream? Well quite a few people but that's beside the point. If you like ice cream, chances are that one of your favourites is vanilla. In fact, in most countries around the world, the biggest selling ice cream is vanilla, which in some places is thought to actually be plain ice cream.

The Americans like it, the Germans like it, the Chinese like it. Even the gelato-loving Italians have vanilla as their favourite. It just tastes good, and it goes with pretty much anything. And of course, you know where vanilla comes from, don't you? Out of a small bottle of dark liquid, right? Well, not totally correct. If you answered "from a vanilla bean" you're getting closer, but it's certainly not a bean. Vanilla actually comes from an orchid, and the vanilla orchid pods are the source of the flavouring (Figure 10.5). Most cultivated vanilla comes from *Vanilla planifolia*, which originated in Mexico, but there are other species native to the South Pacific and the Caribbean that are cultivated. Vanilla orchids grow as vines, usually supported on trees in a plantation. Being tropically adapted, and requiring a high amount of labour in its cultivation, harvesting and processing, vanilla is the second most expensive spice after saffron.

In many cases, when you are eating vanilla ice cream, the flavouring may be a chemically synthesised version of vanillin, as this is much cheaper to make. Of course, you may be one of those foodies who only buys vanilla ice cream with the black vanilla seeds visible in the ice cream, because then you know that real vanilla was used. I am loathe to tell you that the seeds themselves may not be the guarantee you seek, partly because the seeds do not contain any vanillin or other flavours. The flavour components are in other tissues inside the pod but are absent from the seeds.

Figure 10.5   *Vanilla planifolia.*
© Shutterstock.

One of the reasons why natural vanilla generally tastes better than chemically synthesised vanillin is that there are up to 200 other flavour components in natural vanilla. Vanilla experts are like wine judges, they can taste the *terroir* from which the vanilla originated, and for this reason, the vanilla from Madagascar is often considered the finest tasting – although it should be noted that they are growing the vanilla orchids of Mexican origin.

While in Denmark in 2017, I was a guest of Professor Birger Lindberg Møller, a plant biochemist and wonderful man who has been instrumental in not only the discovery of many plant biochemical pathways, but in helping almost countless young scientists develop glittering careers all over the world. Birger now heads the Centre for Synthetic Biology at the University of Copenhagen. He reckons that synthetic biology is third-generation genetic engineering. The research he is leading involves the introduction of complete modules or biosynthetic pathways, not single genes. He also believes, adamantly, that for synthetic biology you need inputs, which must lead to outputs for the benefit of human society. Among other things, Birger's group has been working on cyanogenic glucosides for over 40 years. He first elucidated the production of cyanogenic glucosides in sorghum while a postdoc at UC Davis. He returned to Denmark, and after a stint at the Carlsberg Research Institute (hmm, beer) he gained an academic position at the Royal Danish University for Veterinary and Agricultural Science (KVL). He continued working on sorghum. The cyanogenic glucoside made by sorghum is a compound known as dhurrin, although farmers sometimes refer to prussic acid, another name for cyanide. Dhurrin is synthesised and stored in young or drought-stressed sorghum tissues, frequently at levels that are toxic to livestock. Every year in Australia, stock losses of beef and dairy cattle can be attributed to grazing on sorghum or eating sorghum hay that contains toxic levels of dhurrin, which generates cyanide.

Many other plants are cyanogenic, meaning they make cyanide-containing compounds, most commonly cyanogenic glucosides. Strangely, we still don't really know why. Theories abound as to their utility as forms of insect resistance or nematode resistance, and it was originally thought that if plants can be produced without cyanogenic glucosides (acyanogenic plants), these plants may produce more biomass and be more suited to human or animal consumption. Even more curious is the phenomenon that some plants, like sorghum, do not appear to have acyanogenic versions within the species. Others, like some of the clovers, segregate for the genes that produce cyanogenic glucoside, such that some varieties produce cyanide and others don't. Even more curiously, within some genera, like *Sorghum* and *Eucalyptus*, there appear to be cyanogenic species and acyanogenic species. And to make matters even more confusing, sorghum's closest relatives among the cultivated plants, maize and sugarcane, are acyanogenic.

Sorghum is in the top 10 staple food crops in the world. One of the other top 10 staples, cassava, is also a cyanogenic crop species. Like sorghum, there does not appear to be any acyanogenic varieties within the cultivated species. Cassava (which is sometimes known as tapioca) produces a cyanogenic glucoside called linamarin, as well as some lotustralin, both of which are toxic to humans.[6] It is also toxic to rats and other vermin, which may cause post-harvest losses during storage of the starchy tubers, so this could be one of the reasons for cassava's popularity as a staple. If cassava is not prepared properly, the levels of linamarin could be toxic, or if not toxic, can lead to chronic health problems, such as goitre and ataxia (nerve problems) leading to poor balance and other issues of uneven gait. Hence before it is cooked, the cassava needs to be ground up and spread on the ground with exposure to sunlight, or alternatively soaked in water to leach out the linamarin.

In the 1990s, Birger was able to secure funding from DANIDA, the Danish International Development Agency, to use GM approaches to reduce or totally prevent the synthesis of cyanogenic glucosides in cassava. This would potentially have the benefits of creating non-toxic cassava, overcoming many health problems. What could not be predicted was what, if any, effects the lack of cyanogenic glucosides would have on traits such as insect resistance and plant growth. Given that cyanide (HCN) is a sink for nitrogen, it was hoped that without the biosynthesis of cyanogenic glucosides, the excess nitrogen might lead to improved plant growth and potentially yield. This was a first step into the unknown world of synthetic biology in plants. The team needed to knock out at least one of the enzymes involved in the biochemical pathway for cyanogenic glucoside biosynthesis.

As Birger said, the application of technologies "to the benefit of society" was the stated outcome. Cassava with no toxins and higher productivity. What could possibly go wrong?

The Møller lab used RNAi technology to downregulate some of the key genes encoding enzymes involved in the synthesis of cyanogenic glucosides. They had some collaborators, including cassava breeders in Nigeria. Using the best technologies then available, they were able to produce cassava lines with much reduced levels of linamarin. Linamarin levels in leaves were down to 1% of the normal level, while in the edible part, the tubers, linamarin was 10% of the normal level.[7] This was very good news. The bad news was that cassava just does not grow very well in Denmark. Not surprising really. It's also pretty difficult to grow hydrangeas and tulips in Nigeria. So the obvious thing was to do some field trials in Nigeria. There were a few little problems. For example, when the lines were transformed and taken through tissue culture, it was very difficult to regenerate roots on the plantlets. For the record, if you want a plant to grow in an acceptable manner, roots are a mandatory requirement. There was also some evidence, both under the low light conditions of Denmark and under artificial light, that not being able to make these cyanogenic glucosides was actually deleterious to plant growth. Subsequent work has shown that among the 2500 species of plants that produce cyanogenic glucosides, there is a huge diversity in the role they play in the plant.[7] The counter-intuitive finding that some pathogens actually prefer to infect cyanogenic plants was one, as they can use the HCN as a food source. In addition, there is some evidence that these compounds may be useful as sunscreens under tropical conditions. Yet all of these things were impossible to test in Denmark at 55°N.

Testing in the tropics was required, and given that the cassava-growing regions of Nigeria are 5–10°N it was the logical thing to do. After a number of attempts to gain permission to undertake field trials in Nigeria, the project was terminated – which means we will never know whether these plants were likely to yield well, and have good insect resistance, or were they going to be poor little shrinking violets, unable to deal with the harsh tropical sun. So what was a first-generation effort at synthetic biology in plants was unable to proceed to the point where any decision could be made as to the efficacy of the technology.

To compound the whole issue, anti-GM activist activity was stepping up in Denmark. I suppose one could say that if the activity of activists wasn't stepping up then the activists may not be living up to their job description. A number of GM field trials were destroyed. A glasshouse at KVL was burnt down, and by around 2010 (when Caixia Gao left DLF Trifolium in Denmark), most of the GM crop work in Denmark had closed down. The irrepressible Professor Møller was starting to feel the weight of repression, especially on his team's application of the genetic technologies "for the benefit of society". It was something along the lines of "winter is coming", and Birger moved his science indoors. So . . . back to vanilla.

Birger told me that the vanillin biosynthesis pathway consists of five genes. You can drop that entire pathway as a module into any eukaryote, like yeast. As we've already discussed, there are around 200 flavour compounds in natural vanilla. However, the key compounds are vanillin in its derivative, vanillin glucoside. The combination of vanillin plus vanillin glucoside is more acceptable, dare I say flavoursome, to sensory panels than vanillin alone. Using a synthetic biology approach to develop vanilla flavours makes the vanilla flavour more realistic, to the extent that the yeast may actually add a certain extra flavour component on its own. So what's the benefit? For a vanilla plant, it takes at least eight months to get the vanilla pod to maturity. Making vanillin glucoside in a synthetic biology yeast culture takes eight days, and most sensory panels cannot tell the difference. As Birger told me, people don't really have objections with closed systems where the genetic modification is not out in the environment. In addition, the paperwork is considerably more straightforward than going through the process to grow synthetic biology plants on a farm.

So, is that where modern society wants to end up? Is it better to use energy from the power grid (hydro, wind, solar, gas, biogas or dirty brown coal) than that horrible stuff from the sun to grow a plant outdoors? You're not going to be surprised when I proffer the opinion that from a sustainability viewpoint, I think natural energy from the sun has serious benefits. However, if it is cheaper and more energy-efficient to produce "natural" food (vanillin) or fashion (indigo) components in an industrial setting, then maybe I have to revisit my prejudices. If you can produce the highly desirable vanilla flavour in a controlled indoor situation in Europe, then maybe that is better, more stable and more environmentally benign than to suffer a world shortage because of drought in Madagascar.

You may be aware that from 2016 to 2018 there has been a devastating long-term drought across Southern Africa. Most of the media coverage settled on the fact that Cape Town, a city of 4 million people, may soon run out of water. What is usually ignored (other than in Africa) is the effect that this drought is having on agricultural productivity and livelihoods across the region. Some of the world's poorest subsistence farmers live here. The commercial farmers are certainly feeling the pinch too. This is resulting in shortages of food and local price rises. One of the most high-profile of these is vanilla, because it is a high-end, export-oriented business. Madagascar produces 80% of the world's vanilla. In 2013, vanilla pods cost €42 per kilo. As the drought took hold, prices skyrocketed to €340 per kilo. In March 2017, a cyclone hit and destroyed 30% of the crop, and as a result, vanilla pod prices skyrocketed to €485 per kilo in April 2018.[8] Chefs in France have stopped using natural vanilla in their cooking and criminal gangs are operating in Madagascar to steal vanilla pods from farmers' fields because "it's worth as much as gold". Without synthetic biology and GMOs, we would be seeing the price of vanilla ice cream heading upwards. This may lead to (*quelle horreur!*) chocolate ice cream taking over as the world's favourite.

That is very probably the last thing on the minds of the vanilla producers in Madagascar. They have taken to sleeping out in their fields to try to stop their crop from disappearing overnight. Murders of both thieves and farmers are taking place on a weekly basis. Adding to the complexity (and violence) is the existence of a large timber poaching industry, mostly aimed at illegally logging rosewood for the lucrative market in China. Organised criminal gangs (and possibly even members of the government), many who appear to be using the vanilla industry as a means of laundering their timber poaching incomes, are now organising more illegal forest clearing to plant more vanilla because the price is so attractive.[9] All because rich and privileged people like me love ice cream with little black seeds or rosewood furniture. However, we also have to accept that, like synthetic organic chemistry laid waste to the lucrative indigo industry in the 19th century, synthetic biology could destroy the lucrative vanilla industry in the 21st century.

So in summary, these new kids in town, genome editing and synthetic biology, are powerful tools. In research labs, seed and food companies and research hospitals all over the world these techniques are being used to advance knowledge and understanding of humans, animals and plants. We can already alter/edit almost every gene in every organism if we put our collective minds to it. We can already strip down biological processes to their bare components, and using synthetic biology tools we can put them back together in different ways to make new products or make more products in more time- and energy-efficient ways. Industrial Revolution 4.0 is upon us, because shortly after the knowledge advance comes the application of that knowledge. As Richard P. Feynman said:

> *"The most obvious characteristic of science is its application: the fact that, as a consequence of science, one has a power to do things."*

"Doing things" is of course one part of the equation. For a scientist or engineer, it's the part that gets us out of bed every morning. Getting you, dear reader, to come along for the ride and be convinced that this is good for you, good for the planet, and not harming somebody else is a more monumental issue. Between Ross Cloney's philosophical and mechanical arms there is a head that is concerned about this. As he put it to me:

> *"Synthetic biologists must not make the same mistakes as the molecular biologists did with GM crops in the 1980s and 1990s."*

REFERENCES

1. R. Carlson, *et al.*, *Nat. Biotechnol.*, 2016, **34**, 479–481.
2. E. T. Champagne, K. L. Bett-Garber and M. A. Fitzgerald, *et al.*, *Rice*, 2010, **3**, 270.
3. L. M. Bradbury, *et al.*, *Plant Biotechnol. J.*, 2005, **3**, 363–370.
4. J. Cohen, *Science*, 2017, DOI: 110.1126/science.aa10770.

5. T. Nakano, A. W. Eckford and T. Haraguchi, *Molecular Communication*, Cambridge University Press, 2013.

6. K. Jørgensen, S. Bak, P. K. Busk and C. Sørensen, *et al.*, *Plant Physiol.*, 2005, **139**(1), 363–374.

7. B. L. Møller, Functional diversifications of cyanogenic glucosides, *Curr. Opin. Plant Biol.*, 2010, **13**(3), 337–346.

8. France24, 20 April 2018.

9. J. Watts, The Guardian, 31 March 2018.

CHAPTER 11

For a Better Day

"We are made wise not by the recollection of the past, but by the responsibility for our future."

George Bernard Shaw

"It's tough to make predictions, especially about the future."

Niels Bohr

There are times when decisions need to be made about what sort of future we want for the world. That time is today. It is also every day for the rest of your life. What sort of future world do you envisage? Does your vision for making the world a better place include the wanton destruction of orange petunias? Does it include child labour involvement in farming? Does your vision include millions of children developing blindness because they are vitamin A deficient? I'm guessing not.

I would also be willing to bet that if you harbour feelings of doubt or even downright opposition towards GM plants, animals, foods and fibres, that you nevertheless find blind malnourished children and child labour abhorrent too. Especially if you are an anti-GMO activist, because what you want is a better, healthier and cleaner future for yourself and the planet. And so do I. So I'm sitting here wondering right now – where did the cotton in my shirt come from? How come my petunias aren't

Good Enough to Eat? Next Generation GM Crops
By Ian D. Godwin
© Ian D. Godwin 2019
Published by the Royal Society of Chemistry, www.rsc.org

orange? Maybe I need more vitamins because I'd love to feel 25 again. Yes, agreed, I have a grasshopper mind. Let's just focus for a minute, or better still, until we get to the end of the book. We're almost there.

Golden Rice, the rice that is a golden colour because it contains vitamin A, is still a focus for anti-GM activists. They cannot abide the release of a GM plant. In some ways, Golden Rice is a harbinger of change driven by GM and synthetic biology. A naturally-occurring plant biochemical pathway has been targeted to synthesise beta-carotene in the grain. That grain, once cooked and eaten, can go a long way towards providing the daily requirement of vitamin A for a child. Golden Rice is about to be deployed in Bangladesh. The World Health Organization estimates that 250 million children are vitamin A deficient. Without that vitamin A, up to 500 000 children go blind every year, and half of those die within a year of becoming blind. Yet the anti-GM activists (calling themselves Stop Golden Rice) are still meeting in luxury hotels in places like the Philippines (Mark Lynas, April 2018, Cornell Alliance for Science) to prevent the release of this life-saving rice. Is that the sort of world you want to live in?

You will recall the Bt cotton debacle in Burkina Faso. Monsanto took shortcuts with the breeding and it upset the usual high-quality marketplace that the country was used to playing in. It wasn't the fact that the cotton was GM, it was the fact that the fibre length (staple) was shorter than the older conventional varieties by 0.8 mm. The farmers loved the cotton, but the marketers were not fans, leading to the government stopping the cultivation of the Bt cotton. What have been the outcomes of this decision to phase out Bt cotton in 2015?

Firstly, Burkina Faso is no longer the number one cotton producer in Africa. That honour has gone to Mali since 2017. At its peak, Burkina Faso was producing over 800 000 tonnes of cotton. In 2016/17 this dropped to 683 000 tonnes and then fell further the next season to 563 000 tonnes. Secondly, farmer incomes have gone down because many have been forced to go back to the old practice of spraying insecticide six times per season, which is a major burden on net income. I think we can also agree that this is a major burden on the environment too. Further to that, there is evidence that farmers are returning to the practice of using their children to do much of the additional work required, including spraying. An article published by the Cornell Alliance for Science[1] claims that as profitability has gone down, children have been kept out of school to tend the crops. This may seem a long bow to draw, and it will be some time before the evidence is there as other than anecdotal. However, it is noteworthy that a team of agricultural economists at Oklahoma State University published a study on the effects of growing cotton on school attendance in Burkina Faso. Their results showed that as cotton was introduced to a new area, the average level of farm wealth increased, which led to a higher female school enrolment and a decrease in the use of child labour in cotton-producing areas.[2] However, this was measuring the impact of new cotton-growing areas, not specifically the introduction of Bt cotton, hence the interpretation is not about Bt cotton but cotton incomes. They concluded:

"The results suggest that the income effect from cotton adoption might have been larger than the wage effects from girls, hence the overall positive impacts on school enrolments for girls."

All we can conclude here is that profitability in a low-income society like Burkina Faso is key to societal change and development. The head of the Union of Cotton Producers in Burkina Faso, Francois Traore, says that farmers are lobbying as hard as they can to be able to grow GM cotton again. One farmer from the Hounde district told the Cornell Alliance for Science that in the last year he grew Bt cotton his income was $5370. The very next year when he grew conventional cotton (because of the ban on Bt cotton) his income fell to less than 20% of that. Is that the sort of world you want to live in?

Our final entry for the "Is that the sort of world you want to live in?" would be hilarious if it wasn't so downright ridiculous. It starts in Helsinki. The highly respected journal *Science* wrote about it under the headline: "How the transgenic petunia carnage of 2017 began" (Kelly Sevick, 24 May 2017). Wow! How did you not remember this? Who did these petunias kill? How did the world's governments hush it up? It must have involved collusion with Russia to keep one this quiet. Is Finland now full of zombies who were attacked in the great petunia carnage? No wonder they didn't do so well at the 2018 Winter Olympics.

Teemu Teeri is a Professor in Plant Breeding at the University of Helsinki. In 2015 he walked past the main Helsinki train station and noticed some bright orange petunias. He knew that this was an unusual colour for petunias, and to add to the intrigue, he was aware of research 30 years earlier involving GM approaches to producing new flower colours in petunias (Figure 11.1). Teemu wondered about these petunias so, as interested plant scientists (as well as horticulturists and my mum) have done for time immemorial, he took a little cutting and secreted it in his bag. His analyses demonstrated what he suspected. These plants were GM petunias and contained a maize gene to alter the colour of the flowers, originally from research the Max Planck Institute in Cologne had published in *Nature* over 30 years ago. Further analysis by a group of scientists

Figure 11.1 A variety of petunia examples including: BigDeal Freaky Fuchsia, Crazytunia Terracotta, Hells Bells and Perfectunia Red Improved. Photo by Christian Westhoff Vertriebsges, mbH.

from Vienna and Berlin[3] demonstrated that all these orange petunias, shown to be growing all over the world, could be traced back to one transgenic event. The research was led by Peter Meyer and Heinz Saedler at the Max Planck Institute in Cologne.[4] The original orange petunia was expressing a particular version of a gene for the enzyme dihydroflavanol-4-reductase. The gene was taken from maize, an allele known as A1. These petunias had taken over the world. At least three seed companies from Israel, the Netherlands and Germany had somehow got hold of these interesting lines, not knowing they were genetically modified, and produced new and different orange colours with cultivar names like Electra Orange, Viva Orange and Salmon Ray. Evidently they were very popular, because they were found in North America, across Europe and as far away as Australia. It's uncertain when the first orange petunias hit the market, but there is evidence that they were giving many people joy (as is the main purpose of home and public plantings of flowers).

What happened next resulted in carnage. Teemu told one of his former PhD students who now had a job at the Finnish Board for Gene Technology, and very soon, the carnage began by government order. In late April 2017, the Finnish regulator known as Evira issued a "seek and destroy order". They listed eight petunia varieties, including the romantically named African Sunset and Orange Morn, to be destroyed. Their main message, as well as "ALL ORANGE PETUNIAS MUST DIE" was "THE ORANGE PETUNIAS DO NOT CAUSE ANY RISK TO PEOPLE OR THE ENVIRONMENT". That's what makes me alternate between hysterical laughter and seething anger. Effectively here was a government regulator saying "REMAIN CALM AND CARRY ON DESTROYING PETUNIAS". The lunacy spread. Soon there were similar proclamations elsewhere in Europe, then across the Atlantic to the USA, where the orange petunias were even spotted on university campuses. Then the Australian regulator sent out a similar message, soon to be followed by New Zealand Biosecurity. They tested a lot of petunias and made a very long list on their "seek and destroy order". If you've ever watched an All Blacks rugby match and seen the haka, I think we can agree that the haka is a lot more frightening than petunias called Crazytunia Cherry Cheesecake and Raspberry Blast (although

Hells Glow was also on the list, which may have been the exception that proved the rule). These petunias were out to get us and it was us or them.

So sadly, African Sunset petunias, an All-American Selection in 2014, didn't even get to sail off into the sunset. Before the sun set it was crushed, buried, burned, shot (I kid you not) or sent to landfill, sometimes all at the same time. Or was it? Have a little look on your favourite web browser and type in African Sunset Petunia. Perhaps the sun has not set on the maize A1 gene in orange petunias? I'll put in an order and see (Dear Office of the Gene Technology Regulator – this is a joke, I would never contemplate such an action).

Taking a step back, while the reaction of regulatory bodies was predictable, it was also an incredible lost opportunity. Yes, we all accept that government bodies do not use logic or even necessarily do what is best when it comes to policy and the enforcement thereof. Does your government's policy match its actions when it comes to such things as recreational drug use, immigration, corporate tax or property development? Just asking. Yet with these petunias, one thing all government regulatory systems agreed on was that they had no potential to cause any sort of harm. Yet they still had to die. Here, to my mind, was a perfect opportunity for a government agency to say something like: "These petunias have shown popularity and have not harmed anybody or anything. Hence here is a demonstration that GM plants are no more likely to cause health or environmental issues than conventional plants, however we request that they be labelled as GM plants at point of sale and please enjoy." That's the world I want to live in. A world in which logic prevails. Alas, I also understand why this didn't happen.

1. When logic rears its head that seems to say that regulation is getting it wrong, regulators (bless their hearts) go into a particular mode that is culturally embedded. That public servant mode that replies: "That's what the regulations say. I don't make the regulations, I am just here to ensure they are carried out."
2. Governments govern for everyone in the population, even nutcases. There was every certainty that some nutcase, or group of nutcases, who found each other on the world wide

web would get together to make a class action stating that these orange petunias gave them: recurring headaches/ allergic reactions/bowel cancer/the inability to string two sentences together without saying "like". Or that orange petunias were leading to: the loss of bee populations/ infertility in variegated vicunas/bleaching of the Great Barrier Reef/Muslim immigrants moving in next door.

3. And besides. Petunias are naturally purple. Just like carrots. And corn.

Yes, that's the world we live in. As I will soon come back to, we do live in a world where conflict, real or perceived, sells news stories. Hence if you were to read any form of mainstream media, it is easy to formulate the opinion that there are now two totally different agricultural production systems. It is a binary divide between conventional and organic. Conventional farms all use GM crops and spray deadly toxic chemicals all over the place. They are industrial farms. Organic farms have butterflies and scarecrows, and pretty little rows of multi-coloured lettuces with dewdrops on every leaf, a farmer in dungarees nursing a chicken, and the whole thing is absolutely chemical-free. These are the farms just like in the books your parents read to you when you were five years old. This binary narrative is the bread and butter of many food stories, and is played upon by celebrity chefs and organic wholefood cafes to give you that good earth experience. With pink non-GMO salt.

When it comes to the production of food and fibre, we need to make some changes. Not only to the way in which we produce food, but in the narrative. The narrative of food production is a serious problem. As Stuart Smyth from Saskatchewan told me, it is people like me who are the problem. I talked to Stuart over Skype as we sat in our offices in different hemispheres. Stuart's profile picture is a middle-aged moustachioed man wearing a black t-shirt with a Pink Floyd album cover on it. When he answered Skype, I knew we were going to get along – he was wearing a black t-shirt with a Devo album cover on it. Not that I'm much of a black t-shirt person – I have two. One has a sheep on it and promotes Roquefort cheese, the other my engineer son brought me when he visited NASA in Houston. It was more Stuart's broad tastes in the music from our youths. Stuart holds an industry-funded social science research chair at the University of Saskatchewan. He is actually funded by biotech companies like Monsanto, Syngenta and Bayer CropScience, which is all publicly disclosed on his website. Back to me being the problem. Stuart was not picking on me specifically, rather he was lumping himself and me together with academics working in the field of agriculture.

"As a whole, academics have done a brutal job about communicating what we have done with improving agricultural outcomes and environmental outcomes. The info just doesn't get out there."

Agriculture in the last century has become not only hugely more productive, modern agriculture has also become more sustainable. Whatever your emotional, intellectual and ideological bent, there are some facts that cannot be denied regarding modern agricultural systems.

1. We're now feeding and clothing more people than ever in the history of the planet.
2. Fewer people are required to produce that food and fibre, giving more people time to pursue other activities that improve society. Like teachers, doctors, lawyers, engineers, scientists, nurses, artists, bricklayers, astronauts, professional sportspeople and environmental activists. Of course, it also allows for some people to make society worse, like some doctors, some lawyers, some professional sportspeople, bankers, the mafia, used Citroen salespeople and anti-GMO activists.
3. We're feeding more people per unit of agricultural land than ever before.
4. We have harnessed genetics to make plants and animals more productive, and more efficient.
5. We have improved agricultural management to make plants and animals more productive and more efficient.
6. Crop plants are now more nitrogen and water use efficient than ever before.
7. Crop plants are more resilient to adverse conditions including pests, diseases and climate (hotter, colder, wetter, drier) than ever before.

8. Most of our grandparents spent considerable proportions of their lives with not enough food. That is still the case in many developing economies, yet for a vast proportion of the population, we now have more food than ever.
9. There is a greater variety of foods available to us than ever before.
10. We have reduced the number of times we need to till the soil for weed control, leading to better fertility, richer soil microbial populations and reduced erosion and run-off into waterways and oceans.

However, like all human activities, agriculture has a significant environmental footprint, and it is a constant responsibility and aim of farmers and scientists to reduce the harm of this footprint. All farmers get this, whether conventional, certified organic, GM, biodynamic, permaculture or subsistence. And there is a place for all of them on the planet, although I for one believe that subsistence farming would be wonderful if it became an historical anachronism. Why? Because subsistence farming has been around for 10 000 years or more, and its day is long past. Systems that allow for farmers to produce a marketable surplus are the first step along the path to sustainable development. Cash crops lead to more children going to school, the ability to afford medical care, and have shoes. Better health and better education lead to lower birth rates, and lower levels of poverty.

In the 1800s, an overwhelming majority of the world's population lived in extreme poverty.[5] Even the Industrial Revolution had little effect on this, other than the fact that the rural poor became increasingly the urban poor (Figure 11.2). Right up until the 1950s, more people on the planet were living in extreme poverty than not. The turning point came in the 1970s, and by 1981, for the first time in human history, there were more people not living in extreme poverty that those who were. I'm amazed to think that when I was a university undergraduate in 1981 I didn't know this, but nobody else did either. The number of people living in extreme poverty peaked at 2.22 billion in 1970. At no time before or since have there been more people living in extreme poverty. That number has been dropping ever since. So remember, it was only in 1981 that finally the number of people

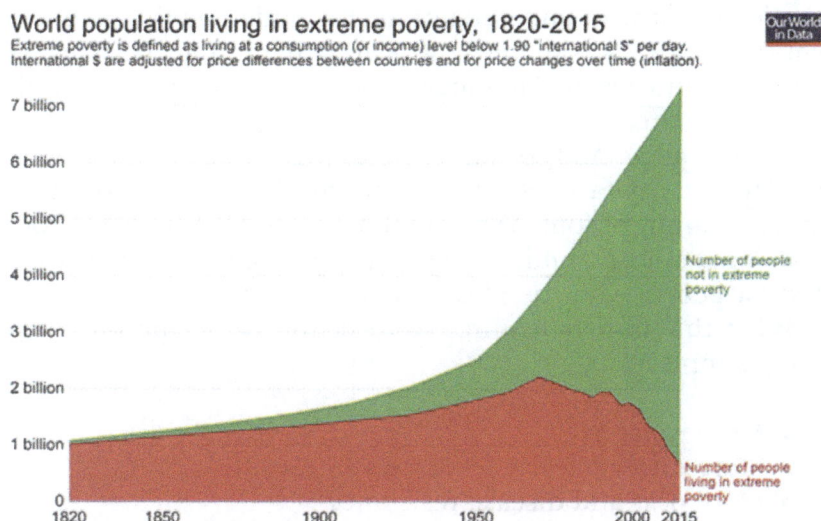

World population living in extreme poverty, 1820-2015

Extreme poverty is defined as living at a consumption (or income) level below 1.90 "international $" per day. International $ are adjusted for price differences between countries and for price changes over time (inflation).

Source: World Poverty in absolute numbers - OWID based on World Bank (2016) and Bourguignon and Morrisson (2002)
OurWorldInData.org/extreme-poverty/ • CC BY-SA

Figure 11.2   The proportion of the world's population living in extreme poverty.
Reproduced from ref. 5 (https://ourworldindata.org/extreme-poverty), originally published under the terms of a CC BY-SA licence.

living in extreme poverty fell, and the first time that the proportion of the population living in extreme poverty fell below 50%. By 2015, the number had fallen to 705 million, and for the first time, less than 10% of the world's population is living in extreme poverty. To me, this represents one of the greatest economic achievements of humankind. The reasons for this are complex, but modern agricultural technologies (as listed above) have been a major contributor, as has the dwindling number of farmers who are subsistence farmers. Most subsistence farmers are extremely low-input farmers, extremely low-output farmers, and nearly all would fit into the category of certified organic, simply because they can afford few inputs other than seed and possibly some animal manure. What can really turbocharge their productivity is relatively simple, but economically unattainable for most.

Imagine this scenario. A maize subsistence farmer on a smallholding in an African country – let's for argument sake say Tanzania. Why Tanzania? I have no idea except it sounds so exotic and African and equatorial. Tanzania has things like big wildlife and Maasai and Mt Kilimanjaro and beaches and Zanzibar and Ernest Hemingway went there and wrote stories. Most of his stories were about shooting large animals and people being unhappy. Add to that the fact that in 2016 Tanzania had a GDP per capita of $879 per annum. So imagine you are an average family of four. You would have a total income of around $3500 per annum, under $70 per week, or about $17 per person. I know people who spend that on coffee per week.

What this farmer requires to turbocharge her productivity is fairly simple:

- More water
- More nitrogen
- More pest and disease resistance
- Better weed control

The average subsistence farmer has no irrigation, no access to synthetic nitrogen fertiliser, no genetics or pesticides for pests and diseases, and a hoe (the weed control implement).

Now imagine a scenario where a farmer has access to a maize cultivar that is drought tolerant and has insect resistance, can control weeds and can afford a bag of fertiliser, something cheap like urea. I will now confess that it saddens me terribly to think that a farmer cannot access these things. And here am I, the privileged white man, on my little urban 400 square metres growing heirloom tomatoes on my home-produced chicken manure. Don't ask me how much input goes into each tomato because I don't know, and the possums and king parrots eat most of the tomatoes anyway. First-world problems indeed. Now back to our Tanzanian subsistence farmer.

Mark Cooper, a quantitative geneticist, and I did our PhDs in labs next door to one another. We started our careers as academics on the same day in 1990 at the University of Queensland. We had labs next door to one another, and worked together seamlessly in teaching and research. After 10 years he got poached by a little seed company in Johnston, Iowa, known as Pioneer HiBred. After another 18 years he's finally back at UQ as a professor. I recently had the privilege to see him at a conference talking about maize (although he's been in the USA so long he calls it corn) improvement. I have to say, he blew the audience away with his talk on AQUAmax corn hybrids, developed by his team in Iowa and now being grown on over 9 million acres across the USA. AQUAmax maizes were developed to be more water use efficient and drought tolerant, and were first released in 2011. There are now almost 70 different hybrids available to farmers, with this technology and combinations of GM herbicide tolerance and Bt insect resistance. Now, imagine a world where it didn't take many millions of dollars to get a GM trait approved for cultivation (I know, I'm dreaming). Then imagine a world where GM activists were not travelling all over the world to whip up fear of GM crops (yep, definitely dreaming). Then imagine our subsistence farmer in Tanzania. What if she could get a microloan to buy this AQUAmax maize? Then what if she could get some nice cheap and safe herbicide to get rid of the weeds on her land (maybe something like glyphosate?). Then imagine if she could get some cheap urea fertiliser to apply to her land (I know, I'm stretching the imagination). So, our subsistence farmer sows the maize in the fertilised land, sprays for weeds four weeks after emergence, the normal corn borers try to eat her plants and cobs and die. The maize survives the usual drought period, and our farmer harvests the crop, which has not been decimated by insects and had to compete for nutrition and water with the weeds.

I don't know about you, but I'm getting attached to this resilient and resourceful lady, so I'm going to call her Irene. Irene harvests the maize and finds she has doubled or trebled her yield because of the improved genetics and fertiliser. She has enough to feed her family with this maize, which has no fuminosin or aflatoxins because of the borer resistance, and she has half her harvest left over. She can sell some of the surplus, then has enough money to buy some chickens for eggs, and eventually meat (more protein, more vitamins). She can feed the chickens with some of the surplus maize. She can send her daughters to school, and get them vaccinated, and buy them shoes, which reduces the chance of them getting hookworms and jiggers. She can buy some vegetables at the market to add some diversity to the diet. Irene can give some of the maize to her husband to feed his pig. With better nutrition, and a little help from the neighbour's boar, his pig can have piglets, and suddenly the farm has gone from subsistence to being part of the local economy. Soon

Irene and her family will join the >90% of the world not living in extreme poverty. Maybe, just maybe, one of her daughters will get enough education to become a schoolteacher, and maybe the other one will get even more and become an agricultural scientist. Well OK, the last one is definitely a dream. She'll probably aim at doctor or engineer or vet.

Yep, dreaming. For this scenario to be remotely possible, there are three requisites that seem almost unattainable with the present world framework. These are:

1. Anti-GMO activists decide to educate themselves on the science and begin to understand the societal and environmental benefits of GM crops. Then they stop telling some of the most resource-poor and least educated farmers on the planet that GM seeds are toxic or a symbol of white colonialism.
2. Seed companies can make their GM crops available in the same way as they make any new variety developed using conventional means. This one may be achievable with genome editing techniques.
3. Farmers can access micro-finance to enable them to buy technology, in the form of improved seeds and fertiliser. Improved seeds are one of the best delivery tools when it comes to agricultural innovation.

It does seem unlikely doesn't it? All we need is a total re-imagining and repositioning of government policy, a willingness for dialogue between warring parties, and media to retake their legitimate position as the Fourth Estate. The historical purpose of the Fourth Estate has been to represent the public interest. In 1787, Edmund Burke called the Reporters' Gallery in the British Houses of Parliament "The Fourth Estate" and more important than the other three (Legislative, Executive and Judiciary). The purpose of the Fourth Estate is to represent the interests of the people such that business or political elites do not abuse their power, whereas in the current world of news on demand, we seem to be falling into a sad state of affairs where "journalism" represents the vested interests of big business, Russian interference or the Kardashians and other people whose names start with K (KGB, KKK, Kanye). I hope that in the previous chapters

I've demonstrated to you that over the past 20 years, the benefits of GM plants represent an instance where the national interest and the public interest coincide.

To continue the dream theme, I'll reiterate my position that the future of sustainable agriculture is to combine the best components of conventional and certified organic agriculture without the ideology. I'll be the first to say that I'm not the only person to make this argument. That argument has been put most elegantly by Pam Ronald and Raoul Adamchak in their 2008 book *Tomorrow's Table* (Oxford University Press). Pam is a Professor in Plant Pathology at the University of California, Davis and Raoul has been a certified organic vegetable producer for over 25 years. Pam has produced a number of GM rice plants with disease resistance and submergence tolerance. She and Raoul have been married since 1996, and their book explores the complementary nature of the power of GM crops and organic production for future sustainable food production. An updated version was published in 2018.

For the remainder of this chapter, we will explore some of the latest results of modern genetics that could produce exceptional outcomes for food and fibre production, while not merely safeguarding the planet, but creating wonderful benefits to the environment.

You've already met the youthful PhD student, Luis Herrera-Estrella, who was central to the race to develop the first transgenic plants in Ghent with Marc van Montagu and Jeff Schell. His youth is still there in his enthusiasm for the future of agriculture, yet of course he is now a very senior scientist heading his own research institute in Guanajuato, Mexico. In fact, Luis returned to Mexico within a year of completing his PhD in 1986, because he felt he had "the possibility to really make a change – something completely new to make life better for Mexican agriculture". He also told me that he felt he wasn't clever enough to keep up as a scientific leader in Europe, which I do not for one minute accept.

When I saw him at a conference in the USA in 1992, he gave an amazing talk about the development of beans resistant to the pathogen *Pseudomonas syringae*, using a technology he called "TacoSavr". For a number of reasons, it never saw the light of day, and a major component of that was what he saw as the misguided and unjustified backlash against GM crop plants.

He was quite discouraged by this, and told me: "I sought refuge in basic science". Among other things, he became interested in the issue of phosphorus uptake and root architecture. Along the way he also took part in research to sequence the pepper genome, the popcorn genome, and the genome of a carnivorous plant. However, it was the issue of phosphorus and plants that really started to take up a lot of his time, and his desire to do something good for Mexico. We all know that in the majority of agricultural systems, nitrogen tends to be the most limiting element to growth and productivity. This does tend to change on acid soils, however. Soils are deemed acidic if the pH falls below the critical level of 5.5. In Australia, about 50% of agricultural soils are deemed to be acidic. Worse still, almost all the soil in equatorial Africa, Brazil, the eastern seaboard of North America, most of Mexico, and most of northern Europe is acidic. There are many and varied problems associated with acid soils but the two factors most limiting to plant growth are that phosphorus become unavailable and aluminium ions reach toxic levels.

Little bit of chemistry? If not, skip the next two paragraphs. But this is very simple chemistry so try and stick with me. Most of the phosphorus (P) in the soil is bound very tightly to soil particles. Plants take up P as anionic phosphates, and hence the standard fertilisers all include bio-available phosphates. Once applied to soils, a significant proportion of phosphate is rendered unavailable, as it is highly reactive with generally abundant cations such as magnesium, calcium, aluminium and iron,[6] depending on the soil type and the pH. This is exacerbated further at depth in the soil, so many plants are adapted to access more P by having shallow, branched rooting systems, forming associations with mycorrhiza to assist with P uptake, and secreting organic acids or phosphatases from the roots to solubilise the P bound to the soil. However, it is a fact of modern agriculture that most of the P applied in fertiliser is never taken up by the plant and eventually will leach out into waterways, resulting in algal blooms and other environmental damage.

Phosphate anions can come in a number of forms, with the most common being the orthophosphate version (PO_4^{3-}). However, there are other anionic versions known as phosphites (PO_3^{3-}), which are reduced forms of phosphate. Luis was intrigued by a 1945 paper stating that phosphite is considerably more soluble in soil than phosphate.[7] Interestingly, phosphite is taken up by plants, but they cannot utilise it. The most stable oxidised form, PO_4^{3-}, is the plant preferred version. However, before plants and the Great Oxygenation Event (when plants invented photosynthesis, or when photosynthesis invented plants) billions of years ago, phosphite was more common. Hence a few bacteria (mostly bacteria that live in low oxygen conditions) have an enzyme called phosphite dehydrogenase, which can then convert the phosphite to phosphate and it becomes a biologically available form of P. Plants naturally lack this enzyme, because they evolved in an oxygenated world. Phosphite has a number of advantages over phosphate in that it is more soluble, has lower reactivity with cations and soil particles, and very few microorganisms in aerobic soils can use it as a source of P.[8,9] Further to that, phosphite does have some agricultural uses in inhibiting the growth of oomycetes, so can provide some defence against diseases like *Phytophthora*).

Luis and his team cloned a gene from a *Pseudomonas* that is a phosphite oxidoreductase. This is an enzyme that can oxidise phosphite into orthophosphate. They expressed the gene in a model plant, *Arabidopsis*, and grew the seedling progenies on a soil that only contained phosphite. The control plants were able to take up the phosphite but could not metabolise it, hence they only grew up to 6 mm and then they grew no further, because they simply ran out of P. Running out of P is not good because it is an essential element for DNA, most of the compounds involved in the energetics of redox reactions, and the phospholipid cell membrane. So if you can't make new DNA, new cells or new energy, things tend to get pretty limiting, pretty quickly. In contrast the GM plants, which could utilise phosphite, grew in much the same way as controls grown on phosphate. Encouraging results indeed, which the team then demonstrated worked just as well in tobacco. The tobacco plants grown only on phosphite as a P source grew in a comparable manner to the wild-type tobacco on phosphate. However, this all begs the question: why would you want to grow plants on phosphite anyway? Some potential advantages have already been mentioned:

- Phosphite is more soluble and hence more plant available, and less likely to bind to soil components.

- There are very few soil microbes in aerobic environments that can utilise phosphite, hence they will not compete for this source of P.
- Hence you can use less phosphite than you would normally apply phosphate.
- Weed control!

Weed control? Well, if you fertilise with phosphate in a P-deficient soil, the weed growth will improve too, hence there is more competition not only for P, but for N, K and other nutrients, as well as water and eventually light if the weeds are tall enough. If you have a crop that can take up and use phosphite, it has a particularly unique advantage over all the other plant species in the field. It will use this source of P, live long and prosper. The other species will founder for lack of a source of P they can use. The Herrera-Estrella team illustrated this beautifully by growing the transgenic tobacco line in the same soil as a weed grass, false brome (*Brachypodium distachyon*), and as you can see (Figure 11.3), on phosphate P the weed seriously outcompeted the crop, whereas on phosphite P, the crop has a significant advantage.[10] Hence the use of this technology has the added bonus as acting as a selective herbicide as well as being a nutrient. This is an elegant and very promising technology, and has the quite considerable benefit of being highly sustainable. Less P is applied to the soil, there is little soil leaching and the need for in-crop herbicide is reduced, if not totally avoided. Luis and his group now have a commercial partner with the plan to get this technology into a number of major crop plants, specifically those important to Mexican agriculture.

As we have already discussed, agriculture has a significant environmental footprint, and one crop with a very large footprint is rice. Given that it is the world's most popular food plant, this is not necessarily surprising. However, as most rice is grown in flooded paddies, it has a very high water use compared to nearly all other crop plants. Another outcome of growth under flooded conditions is that rice produces a significant amount of methane, a consequence of anaerobic respiration of the roots under water. In fact, rice cultivation is the single largest source of methane arising from human activity, accounting for between

Figure 11.3 The comparative growth of transgenic tobacco with the ability to utilise phosphite and a weed grass, false brome. Under no P conditions growth of both is poor. On phosphate the weed outcompetes the tobacco crop. On phosphite only, the weed growth is suppressed and crop growth is dominant.
Reproduced from ref. 10 with permission from Springer Nature, Copyright 2012.

7% and 17% of annual methane emissions worldwide. Methane is a significantly worse greenhouse gas than CO_2.[11] As the roots of rice are in largely anaerobic conditions, when they themselves decompose, or release organic compounds such as sugars, organic acids and amino acids, these are usually converted by methanogenic bacteria, resulting in the escape of methane into the atmosphere. In fact, natural wetlands are the largest source of methane for the same reason. In 2002, it was observed that the higher the grain (and hence carbon) yield of rice, the lower the amount of methane released because there was less carbon in the roots.[12]

Chuanxin Sun from the Swedish Agricultural University in Uppsala decided to express a particular gene from barley in rice. In barley, the SUSIBA2 (sugar signalling in barley A2) gene is a transcription factor which leads to more starch accumulating in the stem. Sun's group expressed the gene in rice under a specific promoter, one normally involved in starch biosynthesis and the branching of starch polymers. This produced rice with significantly more starch in the grain and stem, with lower overall root biomass.[13] They then grew these rices in replicated field trials in China over three years. After all, field trials of rice don't do too well in Sweden. It sounds like they knew what they were looking for, because as well as measuring growth and yield, they also measured methane production and the population of methanogenic bacteria. They found that these rices, as well as being of higher starch content, emitted significantly less methane over the growing season. There were also significantly lower numbers of methanogenic bacteria in the water. Overall, they demonstrated that it was possible to increase the starch content from 77% in the parent line to 87% in the GM lines, with more grains filled. There was no reduction in grain or panicle number, hence overall yield was higher.

Now although this particular advance is specific to rice, it is a significant and important advance. Close to 750 million tonnes of rice is produced every year, and if this can be made more sustainable it is a win–win. Only a small proportion of rice is not grown under flooded conditions, a style of rice cultivation known as upland rice. According to Sun's team, the adoption of these high starch–low methane rices would be the equivalent to shutting down 150 coal-fired power stations or removing 120 million cars from the road (Figure 11.4).

All over the world, applied biotechnology laboratories such as mine are applying genome editing technologies to their crops of interest. While plugging away with our limited funding, we keep a close eye on the literature as groups like Caixia Gao's in Beijing, Jian-Kang Zhu's in Shanghai and Dan Voytas's in Minnesota continue to make incredibly imaginative and inventive improvements in the technical aspects of the use of CRISPR/Cas9 and other emerging technologies. The advances are coming out almost weekly, and it is such an exciting time. The rate of scientific progress is mind-boggling compared to what it was in

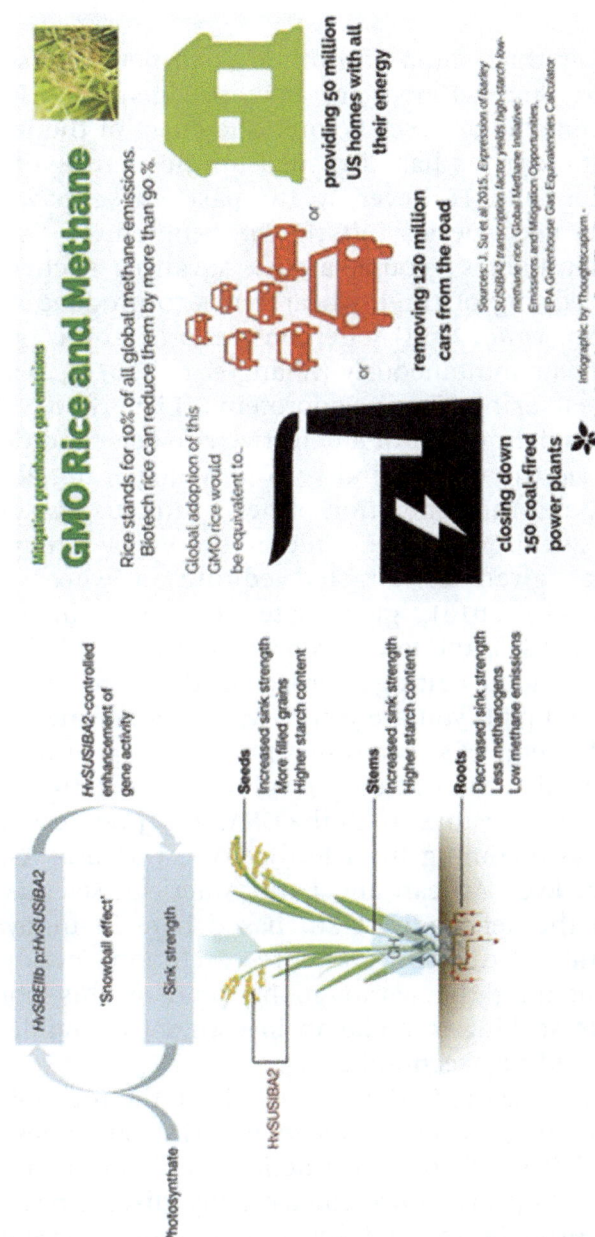

Figure 11.4 Overexpression of a barley gene in rice leads to more grains with higher starch content, and lower methane emissions. If all rice was grown using this GM technology, methane emissions from rice would be reduced by 90%. Reproduced from ref. 13 with permission from Springer Nature, Copyright 2015.

the 1980s when the first transgenic plants were made. In part
this is because the agricultural scientific community is so much
larger.

It's fair to say that China and other Asian powerhouses like
Singapore and South Korea are now producing world-class
scientific outcomes like never before. The effect of the internet
on our communications has also sped up the sharing of information exponentially. However, in the past few years we have
had numerous new species joining the gene edited list from
Caixia and Jian-Kang's groups alone. Advances such as the
simultaneous editing of three wheat genes to produce mildew
resistance (Wang *et al.*, 2014), a demonstration that it is possible
to edit six genes simultaneously (Zhang *et al.*, 2016), DNA-free
editing of wheat using ribonucleoproteins (Liang *et al.*, 2017),
homology-dependent repair of a defective rice gene and delivery
of gene edits *via* the use of viral vectors (Wang *et al.*, 2017) or
nanoparticles, and the generation of new glutinous (waxy) rices
(Zhang *et al.*, 2018). Not to be outdone, the Voytas group from
Minnesota has given us targeted editing in wheat pollen
(Bhowmik *et al.*, 2018), glyphosate resistance in cassava
(Hummel *et al.*, 2017), and low gluten wheat (Sanchez León *et al.*,
2017), all from genome editing approaches (these are not GMOs).
Are you keeping up? If you are you're way ahead of me.

We are on the verge of so many new products appearing on our
food shelves. Unlike the GM crop revolution of the 1990s when
most of the advances came out of the USA, this time we will see a
raft of things also coming from Europe, the USA and Asia, especially China. Even African and Latin American scientists are
getting in on the act. Sadly, I am based here in innovation-
unready Australia. Our main grains research and development
investor is watching the world go by "and at this point in
time .... is not seeking to make an investment in the development of (gene) editing technologies".

So what this is saying is that we now have the technology to
use genome editing (which leads to non-GM outcomes when
done well) to deliver all sorts of genetic advances to major food
and fibre crops. Improved nutritional quality, disease resistance,
adaptation to drought and salinity, and even gluten-free wheat!
The promise of genome editing is not just a promise. It's here
now, and plant breeding is progressing at a rate never before

seen. We are in the midst of Industrial Revolution 4.0, and the agricultural world is in for major changes such as we have not seen before. New genetics and digital agriculture are in the midst of combining to make food and fibre production more sustainable and more efficient in ways that nobody can yet imagine. Most of us in the agricultural science world cannot contain our excitement, and just as the digital revolution has irreversibly changed communications, it is society and our governments who are scrambling to keep up. Nowhere do they scramble more than when it comes to issues of animal welfare and genetic manipulation.

Gene edited agricultural animals are already upon us, as already explored when it comes to polled (hornless) cattle (Chapter 10). But wait, there's more. Lots, lots more. For the sake of illustration I will use pigs, because as previously established, everything goes better with bacon, even for vegans.

One of the most devastating porcine diseases is a virus, porcine reproductive and respiratory syndrome (PRRS), which causes severe breathing difficulties in piglets and infertility in sows. The disease is estimated to cost the pork industry €1.5 billion every year in Europe alone. Scientists at the Roslin Institute, part of the University of Edinburgh, with a company called Genus, have applied CRISPR/Cas9 technology to produce pigs with resistance to PRRS.[14] The virus targets a cellular protein known as CD163, which the Roslin/Genus team knocked out using CRISPR/Cas9. The resultant piglets were resistant to two known sub-types of PRRS. The genetics will be made available through Genus, a company specialising in cattle and pig genetic improvement, who currently have a market for their genetics in 70 different countries.

Another recent breakthrough has been the use of CRISPR/Cas9 technology to improve meat quality from pigs. A group from the Chinese Academy of Sciences in Beijing has produced some CRISPR-edited pigs to express a gene variant found in mice. This gene, known as UCP1 (uncoupling protein 1) gives cells the ability to dissipate more heat and burn more fat while doing so (Zhang *et al.*, 2017). The resulting pigs had 15–20% less body fat and a higher percentage of lean meat. How did the animals fare health-wise? This simple change actually made the pigs more resilient to temperature changes. The genetic edit led to pigs that were better able to keep their body temperature more stable, and actually made them more cold-tolerant.

Most people are aware of the influenza pandemic that devastated the Earth's population and coincided with the end of World War One in 1918. Military personnel returning from the war to places all over the globe exacerbated the spread of the virus. The influenza, widely known as Spanish influenza, is now widely accepted as an influenza A, H1N1 influenza. It was estimated to have killed 3–5% of the world's population and was the deadliest pandemic in human history. In addition to being deadly, it was not Spanish. The heavily censored press during and directly after the war prevented reports of the virus. It was deemed to be an issue that could devastate morale. Instead, they allowed it to just devastate the population. As Spain was not involved in the war, the influenza outbreak was widely reported in Spanish newspapers (Aj caramba, there was no internet), particularly because King Alfonso XIII was gravely ill with the disease. Interestingly, nobody knows the true origin, but there was huge mortality in both human and pig populations in Europe. Some later analyses reported that the influenza was already rife in Kansas in 1917, and US troops took it to Europe. Others felt they have enough evidence to conclude that the strain arose in Austria in 1917, whereas others came to the conclusion that this particularly nasty strain came to Europe with the almost 1 million Chinese war labourers. Nobody wants to claim to be the source of such pestilence, not unlike Dutch elm disease, a fungal disease which arose in Asia, or the Colorado potato beetle which was first observed on potatoes in Nebraska. What we can conclude is that the disease most certainly did not originate in Spain, and that it crossed the species barrier from either an avian or a porcine origin.

H1N1 is a virus that as well as infecting humans has types which affect both poultry and pigs. It remains a subject of much debate as to whether the original strain of the Spanish influenza outbreak crossed species boundaries into humans from pigs or chickens. What we do know is that the WHO declared a swine flu pandemic in 2009 when a new strain of H1N1 jumped into humans from pigs. There have been recent bird or avian flu outbreaks originating in Asia although many of these are a different strain known as H5N1. The first spread of H5N1 from birds into humans started in Hong Kong in 1997. Avian influenza has killed millions of birds worldwide, but mostly in East and South East Asia. In Vietnam alone, over 50 million domestic birds were killed by the disease or as a quarantine measure. What if we could do something to protect animals from these diseases, and as a result, reduce the risks of the disease "jumping" into the human population?

Using a CRISPR/Cas9 gene activation system, a group at Duke University led by Professor Nicholas Heaton were able to identify a key gene involved in the viral infection and virulence process. The gene is B4GALNT2, which produces an enzyme called beta-1,4 N-acetylgalactosaminyltransferase 2, and it is responsible for producing an antigen found in red blood cells. Overexpressing this gene in cell cultures demonstrated that this prevented infection of every avian influenza virus strain tested, including all sub-types of H5, H7 and H9.[15] The ability to produce genome-edited birds with this change would prevent the disease in many hundreds of millions of poultry birds (chickens, ducks, geese), providing much-needed nutrition and protein to millions of consumers in the form of eggs and meat.

This is just the start. New technologies are going to revolutionise the way we produce food and fibre, and new powerful genetics will drive many changes that we still have not even dreamed about. It doesn't matter whether you are an omnivore, a herbivore, a vegetarian or only eat organic, these new technologies will ensure that the world does become a better place. The footprint of modern agriculture will be reduced, animal welfare and human health will benefit through the prevention of disease, and the production of healthier and more nutritious foods. So just remember, we humans are a diverse lot. You may not agree with how others choose to farm or choose to eat, but if we keep improving food and nutritional security, keep improving the economics of efficient food production while simultaneously making for a better environment, everyone is a winner. Well, except maybe breatharians. No, thinking about it, breatharians will be better off too because they won't get bird or swine flu and will have cleaner drinking water. They will also breathe better air if we reduce diesel emissions by not having to constantly drive up and down the field with pesticides, and I include copper and rotenone in that list, especially because they have to be applied so frequently.

What is in the way, blocking these improvements? We've already established it's me because Stuart Smyth said so. But it's also you. Yes, you, the one sitting on the sofa with your bowl of organic udon noodles and cup of sencha tea. Do you really want the world to be a better place? Or do you just want to feel good about yourself, because you're eating organic noodles? It's OK. I know it's hard to get information these days. Actually, it's not. It's easy to get information. All you need to do is "google" GMO and you will have no end of misinformation and fake news.

The anti-GMO activists are full of advice on how to avoid GMOs and why you should avoid GMOs, much of it totally wrong. Why is that? While doing the research for this book I had the pleasure and privilege to speak with many people. Speaking to thoughtful and energetic individuals like Stuart Smyth (Saskatoon), Alison van Eenennaam (UC Davis) and Urs Niggli (FIBL), I have had time to develop a clearer picture now. Stuart, in his black t-shirt, summed it up really well. In the early halcyon days of genetic engineering (as we called it then), we scientists were just too excited about the science and what we could now do to pay much attention to what the general public thought. We thought what we were doing was great, so why wouldn't everyone with half a brain (or hopefully more)? Then came the Monarch butterflies and StarLink corn in the USA, and European environmental activists asking questions. Initially they had some good questions. What about biodiversity? What about weediness? How about access for farmers in developing countries? What about non-target organisms? Academia said "Good question, we will answer that!". Questions were answered and by 2005, production figures (yippee) and statistics on the reduction of pesticide use (woohoo), and 90% decreases in soil run-off and erosion (yeehaa) were available and scientifically replicated. So activists started to ask more questions. As question after question was answered scientifically, they asked more and more farfetched questions. Vested interests joined in, mostly from the organic food lobby, which was increasingly representing big business. Stories had to be made up about GM crops and foods being toxic, the tools of industrial agriculture, and the intrinsically evil multinational seed companies forcing farmers to buy their seed – all in the face of the reality of facts. Big organic agriculture has become industrial. No person has ever died or been harmed by a GM crop or food. Farmers the world over have a choice of where they buy their seeds, except for the subsistence farmers who would love to be able to buy some little biological packages of technological advancement, yet cannot. They cannot because they either cannot afford to buy any seed, or their governments have succumbed to the lies pedalled by groups such as Greenpeace and Friends of the Earth.

As Stuart told me, the anti-GMO activists, especially the high-profile groups like Greenpeace and Friends of the Earth, have painted themselves into a corner. Urs Niggli from the Organic Agriculture Research Institute said much the same. He soon realised that as he stood in front of activists proposing that perhaps we should embrace CRISPR technology, the activists became agitated. Urs was challenging their very livelihoods. Activist organisations have continued to raise campaign dollars, pounds and euros to prevent the technological advances of GM crops from spreading and showing the world what can be achieved. They have now put themselves in a predicament. If they admit, à la the Mark Lynas *mea culpa*, that they were wrong or misinformed about GM crops and foods, and changed their public opinions, they may well be set upon. The public, including their followers with a strong social conscience and environmental ethos, could well turn against them. The public will resent that they were lied to, and this will cause doubts as to their other campaigns. Important campaigns such as gaining traction in the fight against climate change, preventing wanton destruction of habitats and forests, and cleaning up the oceans from plastics will leave people wondering, "Have I been told the truth?". Above all, they have to stop technologies like Golden Rice because they fear that this will be a wonderful success if adopted. They rightly (or wrongly) fear that Golden Rice will prevent childhood blindness and illness, and as a consequence, save many lives. If anti-GMO activists really held the opinion that Golden Rice would be a failure and was not going to save any lives, they would let it go ahead so they could use it as an example. However, they think it will work, and this will open the floodgates for many other GM and gene edited technologies to enter the agricultural marketplace. So they are doing their utmost to make sure it does not happen. If that requires a healthy dose of fake news then so be it. Hence Golden Rice is "toxic", is against the "religious and moral beliefs of minorities" and "farmers should just grow more leafy vegetables". If you are a wealthy middle class European that perhaps makes sense, especially because you can just go to Aldi and buy some spinach. If you are a malnourished subsistence farmer whose number one concern is for you and your family to get enough calories, the aversion to growing less rice (calories) and supplanting it with spinach or cabbage (less vitamin A than Golden Rice and virtually no calories) is understandable. There is also the fact that in the tropics, leafy vegetables are the favourite meal of many butterflies, moths, leafhoppers, aphids and beetles. So pesticide would probably be required too. Some activists, like Greenpeace, have actually changed their narrative from "this will never deliver enough vitamin A" to "nobody knows what adverse effects too much beta-carotene (pro-vitamin A) will have on human health". While it is possible to reach toxicity with vitamin A, nobody has actually ever eaten enough beta-carotene for toxic levels to be reached.

As part of his role as an educator, Stuart Smyth has had his undergraduate classes in Saskatoon indict Greenpeace for crimes against agriculture. The students have had to defend or prosecute the case against Greenpeace. Most of his students are from farming backgrounds, and hence Greenpeace is not one of their favourite organisations. However, if nothing else, his students learn what it is that Greenpeace has against agriculture and some of the techniques they and other anti-GMO activists use to besmirch modern food production. These besmirching techniques will in most cases only work for people who have never been involved in food production.

In conclusion, it is misinformed anti-GMO activists, and the organic lobby groups who have sought over the past 20 years to make GM crops and foods controversial. Organic food companies like Whole Foods Market (now part of Amazon), whose annual revenue is about the same as that of Monsanto (around $15 billion in 2016), are no longer small players. They are large industrial companies with international logistics and marketing operations to match (Amazon is one of the largest companies in the USA, with annual revenue of $178 billion in 2017). For the world to become a better place, we all need to stop pretending that organic farmers are the little guys. We also need to stop pretending that certified organic farming and food leads to the most sustainable agriculture and healthier food. This has been analysed time and time again, and it is demonstrably not the case. By far the biggest problem with certified organic farming is the restrictive practices enshrined within their rules. Many of their restrictions are based on the demonstrable misconception: "If it's natural, it's good". Modern genetic technologies are going ahead in leaps and bounds. As we move beyond the herbicide tolerance and Bt insect resistance paradigm, the promise of not just genomics and GM techniques, but the new genome editing techniques, is simply astonishing. Embracing the best parts of the new breeding technologies will help certified organic agriculture overcome its greatest limitation. This limitation is that even with the best agronomic and land management practices, you simply cannot produce more food and better food on the same land area unless you embrace better genetics and better plant and animal nutrition. Together, using the best farm management practices (especially modern fertilisers), the best animal welfare practices, and the best modern plant and animal genetics, we can produce more safe and nutritious food with less input and lower environmental footprint. If we can get past the artificial binary divide between "conventional" and "organic", resist and call out the fake news, and work together using the best evidence-based science and social responsibility, we can make sure all our food is Good Enough to Eat.

BIBLIOGRAPHY

P. Bhowmik, E. Ellison, B. Polley, V. Bollina, M. Kulkarni, K. Ghanbarnia, H. Song, C. Gao, D. F. Voytas and S. Kagale, Targeted mutagenesis in wheat microspores using CRISPR/Cas9, *Sci. Rep.*, 2018, **8**(1), 6502.

H. A. C. Denier van der Gon, *et al.*, *Proc. Natl Acad. Sci. U. S. A.* 2002, **99**, 12021–12024.

A. W. Hummel, R. D. Chauhan, T. Cermak, A. M. Mutka, A. Vijayaraghavan, A. Boyher, C. G. Starker, R. Bart, D. F. Voytas and N. J. Taylor, Allele exchange at the EPSPS locus confers glyphosate tolerance in cassava, *Plant Biotechnol. J.*, 2017.

Z. Liang, *et al. Nat. Commun.*, 2017, 10.1038/ncomms14261.

S. Sánchez-León, J. Gil-Humanes, C. V. Ozuna, M. J. Giménez, C. Sousa, D. F. Voytas and F. Barro, Low-gluten, non-transgenic wheat engineered with CRISPR/Cas9, *Plant Biotechnol. J.*, 2017.
Wang, *et al.*, *Mol. Plant*, 2017, **10**(7), 1007–1010.
    Z. Zhang, *et al.*, *Plant Cell Rep.*, 2016, **35**, 1519–1533.
Q. Zhang, *et al.*, *Proc. Natl. Acad. Sci. U. S. A.*, 2017, 114, E9474-9482.
J. Zhang, *et al.*, 2018, J. Integr. Plant Biol. https://doi.org/10.1111/jipb.12620.

REFERENCES

1. Cornell Alliance for Science, Assessing the real cost of Burkina Faso's decision to phase out GMO cotton, https://allianceforscience.cornell.edu/blog/2017/12/assessing-the-real-cost-of-burkina-fasos-decision-to-phase-out-gmo-cotton/ [accessed July 2018].
2. H. Kazianga and F. Makamu, *Am. J. Agric. Econ.*, 2017, **99**, 34–54.
3. C. Haselmair-Gosch, S. Miosic, D. Nitarska, B. L. Roth, B. Walliser, R. Paltram, R. C. Iucaciu, L. Eidenberger, T. Rattei, K. Olbricht, K. Stich and H. Halbwirth, *Front. Plant Sci.*, 2018, DOI: 10.3389/fpls.2018.00149.
4. P. Meyer, I. Heidmann, G. Forkmann and H. Saedler, *Nature*, 1987, **330**, 677–678.
5. M. Roser and E. Ortiz-Ospina, 2018, Global Extreme Poverty, 'https://ourworldindata.org/extreme-poverty [accessed July 2018].
6. D. L. López-Arredondo, M. A. Leyva-González, S. I. González-Morales, J. López-Bucio and L. Herrera-Estrella, *Annu. Rev. Plant Biol.*, 2014, **65**, 95–123.
7. R. H. Bray and L. T. Kurtz, *Soil Sci.*, 1945, **59**, 39–46.
8. S. C. Morton, D. Glindemann, X. Wang, X. Niu and M. Edwards, *Environ. Sci. Technol.*, 2005, **39**(12), 4369–4376.
9. A. K. White and W. W. Metcalf, *Annu. Rev. Microbiol.*, 2007, **61**, 379–400.
10. D. L. López-Arredondo and L. Herrera-Estrella, *Nat. Biotechnol.*, 2012, **30**, 889–893.
11. T. F. Stocker, D. Qin, G.-K. Plattner, M. M. B. Tignor, S. K. Allen, J. Boschung, A. Nauels, Y. Xia, V. Bex and P. M.

Midgley, *IPCC Climate Change 2013 The Physical Science Basis, Working Group/Contribution to the Fifth Assessment Report of the Intergovernmental Panel on Climate Change*, Cambridge University Press, New York USA, 2013.

12. H. A. C. Denier van der Gon, M. J. Kropff, N. van Breemen, R. Wassmann, R. S. Lantin, E. Aduna, T. M. Corton and H. H. van Laar, *Proc. Natl. Acad. Sci.*, 2002, **99**, 12021–12024.

13. J. Su, C. Hu, X. Yan, Y. Jin, Z. Chen, Q. Guan, Y. Wang, D. Zhong, C. Jansson, F. Wang, A. Schnürer and C. Sun, *Nature*, 2015, **523**, 602–606.

14. C. Burkard, S. G. Lillico, E. Reid, B. Jackson, A. J. Mileham, T. Ait-Ali, C. B. A. Whitelaw and A. L. Archibald, *PLoS Pathogens*, 2017, **13**, e1006206, DOI: 10.1371/journal. ppat.1006206).

15. B. E. Heaton, E. M. Kennedy, R. E. Dumm, A. T. Harding, M. T. Sacco, D. Sachs and N. S. Heaton, *Cell Rep.*, 2017, **20**(7), 1503–1512.

Subject Index

A1/A2 mating type, *Phytophthora infestans* 218, 220, 221, 226
2-acetyl-1-pyrroline 254–5
activists
    anti-GM *see* anti-GM activists and protestors
    pro-GM 163, 236
ad hominum attacks by activists 236
Adamchak, Raoul 283
adenine (A) 36
Advanta 82–3, 84, 85, 86, 150, 153, 161
aflatoxins 116–17, 118, 281
Africa
    cotton 101–3, 271–2
    maize 115
    Southern, drought 267
Agent Orange 4
agricultural biotechnology (agbiotech) companies 79, 80, 84, 86, 88
    start of revolution (1980s) 79, 80, 84
agriculture and farming
    animal gene editing 291

biodynamic 204–6, 240, 241, 250, 277–8, 282
changes (modern times) 134–6
environmental footprint 278, 286
India, farmer suicides 175–9
organic *see* organic agriculture
prices for products *see* prices
productivity *see* productivity
revolutions *see* revolutions
subsistence 2, 278, 279–82, 295, 296
US (in 1940s) 26
Agrobacterium faciens 40, 42, 44, 58, 65
    crown gall disease 40–1, 43, 45, 48
    plasmid (Ti) 42, 43, 44, 48
Aitken, Liz 121, 127, 128
2S albumin 140, 141
alcohol dehydrogenase genes 38–9

alkaloids, potato 225
allergies (food) 18–19, 140–1
    kiwi fruit 154
    maize 150, 153–4
    peanut 18
    skin prick test 140–1
    soybean 140–1, 155
    tree nuts 140, 141
Alomae–Bobone viruses 70
Amcor and Visy 89–90
Amflora potato 228–9
Amiens (Queensland) 190
amino acids 138, 139
    essential 138, 139
    glyphosate actions 56–7
    in protein synthesis 36
Ammann, Klauss 162–4
ammonia in Haber–Bosch
  process 26, 78, 176
Andes, potato 225
animals
    fat 143, 144, 145
    gene editing 291
    pet, chocolate toxicity 197
antibiotic resistance 31–2, 40,
  46, 49
anti-GM activists and protestors (incl. eco-warriors)
  94, 136–8, 203–4, 231, 236,
  282, 294–7
    allergies 141–2
    Europe 162–7, 227–8, 230
        Denmark 266
        France 166, 228–9
        Germany 53, 164,
          166, 167–8,
          203, 215
        Switzerland 165–6
    Friends of the Earth 55,
      151, 167, 170, 227, 295

Greenpeace 53, 55, 61,
  112–13, 113, 118, 131,
  174, 231, 233, 295, 296
    India 175–6
    Lynas (Mark), and his
      changing opinions
      230–3, 295
    maize and 151–2
    Monsanto and 91–2, 168,
      231, 234
    organics and 203–4
    papaya 131–2, 133–4
    personal attacks by
      235–6
AQUAmax maize 283
Arabidopsis 215, 256, 285
Argentina
    Balcare in 81, 214
    soybean 61, 112, 113, 114
aroma *see* flavour; smell
asbestos 93, 200
asparagine 36, 57
aspartame 93
Aspergillus toxins (aflatoxins)
  116–17, 118, 281
Australia, *see also* Queensland
    banana (and Panama dis-
      ease) 126, 127–9
    biosecurity *see* biosecurity
    celery 191–2
    chemical/agbiotech com-
      panies 81–2
    cotton (and insect pests)
      62–3, 65–6, 67, 103–8
    CSIRO 65–7, 165, 206,
      209, 210
    divisive political argu-
      ments in 207
    Friends of the Earth 55,
      151, 167, 170, 227, 295

Australia, *see also* Queensland
 (*continued*)
    Greens Party 142
    innovation unreadiness
       290
    rice 253, 254
    science communication
       210
    spraying with insecticide
       62–3, 65, 67, 105–6,
       108, 126, 136
    superphosphate and its
       impact 77
    WWI soldiers in France
       from 3–6
autism in India 175
Aventis 86, 149–50, 152, 153,
 154, 181
Avery, Oswald 31
avian flu 292–3

B4GALNT2 293
Bacillus thuringiensis (Bt) 40,
 62–8, 103, 186–7
    cotton 62–3, 65–8, 100,
       101, 102–3, 106, 108, 118
       Burkino Faso 101–3,
          271
       India 98–9
    maize 48, 50–1, 68, 116,
       117–19, 149–50, 179–85,
       203, 228
       and Monarch but-
          terflies 179–85
    toxin 64, 65, 66, 67, 69,
       99, 181, 182, 185, 203
       Cry proteins 64, 65,
          67, 103, 118, 150,
          151, 152, 153,
          154, 181

backcrossing and cotton
 101, 102
bacon 146
bacteria 32
    antibiotic resistance
       31–2, 40, 46, 49
    immune system, CRISPR
       and 247
BAD2 (betaine aldehyde dehydrogenase 2) 255, 256
Baeyer, Adolf von 78
Balcare 81, 214
bananas 122–9
    Cavendish 121, 123,
       124, 126
    Panama disease 122–9,
       155
    potassium 60
bar gene 50
barley SUSIBA2 gene 288
BASF 30, 78, 84, 88, 91, 213,
 214, 215, 226–7, 227, 229
basmati rice 252, 253
Bayer 86
    takeover of Monsanto 88,
       90–1
Beijing, Chinese Academy of
 Sciences 255, 256, 291
Belgium 167
Berkeley (UC), CRISPR 256,
 257, 258, 259
Berliner, Ernst 64
beta-1,4 *N*-acetylgalacto-
 saminyltransferase-2 293
beta-carotene 147, 231, 271,
 296
betaine aldehyde
 dehydrogenase-2
 (BAD2) 255, 256
bialaphos herbicide 50

Bio (label) 194
Bio-Win 83
biodynamic agriculture 204–6, 240, 241, 250
biosecurity and biosafety
    Australia/NZ
        banana and Panama disease 128, 129
        papaya ringspot virus 130
        petunias 274–5
    Europe 166
    Switzerland 164, 165
biotechnology *see* agricultural biotechnology; anti-GM activists and protestors
bird flu 292–3
black Sigatoka 121, 126, 127, 128
Bohr, Niels 208, 270
boll weevil, cotton (*Helicoverpa*) 62, 64, 65, 66, 68, 104, 108
Bollgard (I/II/III) 65, 67, 98, 99, 101, 103
bollworm, cotton (*Helicoverpa armigera*) 62–3, 65
Bond, Jack and Olive 2–3
Bordeaux, grape sprayed with copper sulphate 222
Bordeaux mixture 222, 223, 224, 227
Borlaug, Norman 20, 28–9, 78
Botella, Jimmy 168–9, 170
Bové, José 228–9
bovine growth hormone 148, *see also* cattle/cows
Brazil, soybean 61, 109, 112–15
Brazil nut 139–41

breeding technologies, new (NBTs) 241, 244, 250, 297
Broad Institute 257–8, 259
broccoli 190
brown snake, author encountering 104–5
Bt *see Bacillus thuringiensis*
budworm, cotton (*Helicoverpa punctigera*) 62–3
Burkina Faso 101–3, 271–2

cabbage, CRISPR-edited 244
caffeine 198, 199
Calgene 85, 160
callus tissue 46, 49
cancer (and carcinogens) 200–1
    aflatoxins 116
    Agent Orange components 92
    cervical cancer 94
    glyphosate 200, 202
    tobacco 193–4
carcinogens *see* cancer
cardboard boxes 88–90
Carica cauliflora 71–2
Carica papaya see papaya
β-carotene (beta-carotene) 147, 231, 271, 296
Carson's *Silent Spring* 92
cartels 89–90
Cas9 247, 250, 256, 257, 258, 259, 260, 291
cassava 264–5
cattle/cows, *see also* bovine growth hormone
    horn *see* horn
    sorghum and 263
cauliflower mosaic virus 40, 58

Cavendish banana 121, 123, 124, 126
CDC (Centers for Disease Control and Prevention) 142, 154
celery 189–93, 194, 197
cell cultures, human genome editing 257
cell wall
 Fusarium oxysporum 125
 plant 33, 34, 159
Center for Biological Diversity 183
Center for Food Safety 151
Centers for Disease Control and Prevention (CDC) 142, 154
Central Institute of Cotton Research (CICR), India 99
cereals (cereal grains) 48, 51, 139
 costs over time 87
 paleo diet and 138
cervical cancer 94
Champagne, Elaine (and her group) 252, 253
Charles (Prince of Wales) 157, 177
Charpentier, Emmanuel (and his team) 257, 258
chayote (choko; *Sechium edule*) 6–7
ChemChina 88
chemical companies 74–95
 historical perspectives 75–9
chemical fertilisers *see* fertilisers
"chemical-free" food 14, 254
chemoreceptors 15

children
 autism in India 175
 choco and school milk 7–8
 food allergies 18
 vitamin A deficiency 271
Chilton, Mary-Dell 41, 42, 43, 46
chimeric genes 37, 40, 46
Chinese Academy of Sciences in Beijing 255, 256, 291
Chinese agrochemical companies 88
Chinese food 10
chlamydospores and banana tree 125, 128
chloroplast, protein transport to 59
chocolate 61
 toxicity with pets 197
choko (chayote; *Sechium edule*) 6–7
climate change (incl. global warming) 75, 183, 207, 231, 295
 denial/non-acceptance 207, 208, 231
 methane as greenhouse gas 286–8
Cloney, Ross 259–60, 268
Clustered Regularly Interspaced Short Palindromic Repeats (CRISPR) 129, 243–59, 291, 293
coal *vs* renewables in Australia 207
Cocking, Ed (and his group) 33, 34, 35, 40, 47
cocoyam (taro) and aroid root crops 68–70, 71

coeliac disease 19
coffee 14, 125, 166, 178–9
Cohen, Stan 32, 33, 49
colours, petunias 273–6
common brown snake, author
    encountering 104–5
Commonwealth Scientific and
    Industrial Research Organ-
    isation (CSIRO) 65–7, 165,
    206, 209, 210
companies *see* corporations
    and companies
competition 89
Cooper, Mark 280–1
copper (sulphate/CuSO₄)
    217–18, 224
        plus slaked lime (Bordeaux
        mixture) 222, 223,
        224, 227
Cormick, Craig 66, 206,
    209
corn *see* maize
corn borer 150, 281
    European 68, 116, 117
    Mediterranean (*Sesamia
        nonagrioides*) 118
    South West 117
Cornell Alliance for
    Science 131, 271, 272
Cornell University 48, 72,
    131, 180
corporations and companies
    agricultural bio-
        technology *see* agri-
        cultural biotechnology
    chemical *see* chemical
        companies
    "evil" 92, 233
    multinational 133, 136,
        163, 233–4, 234

productivity and com-
    panies involved in
    1980s 280
costs of agricultural products
    see prices
cotton 65–8, 98–108
    Bollgard (I/II/III) 65, 67,
        98, 99, 101, 103
    Bt *see Bacillus thuringiensis*
    India 98–9, 175, 176,
        177–9
    monocultures 62, 175
    Monsanto and 65, 66, 98,
        99, 100, 101, 102,
        103, 271
cotton boll weevil (*Helicoverpa*)
    62, 64, 65, 66, 68, 104, 108
cotton bollworm (*Helicoverpa
    armigera*) 62–3, 65
cotton budworm (*Helicoverpa
    punctigera*) 62–3
cotyledons 47–8
cover pricing 90
cows *see* cattle; horn
crassulacean acid metabolism
    photosynthesis 169–70
creationist scientists 207
CRISPR 129, 243–59, 291,
    293
crops and crop plants
    in modern agricultural
        systems 277
    spraying with insecticide
        see spraying
    yields, growth 20
crown gall disease 40–1,
    43, 45, 48
Cry (crystalline) proteins 64,
    65, 67, 103, 118, 150, 151,
    152, 153, 154, 181

CSIRO 65–7, 165, 206, 209, 210
cultural identity 9–10
cyanogenic glucosides 263–5
cymoxanil 223, 224
cytosine (C) 36

2,4D 92
Danaus plexippus (Monarch
    butterflies) and Bt maize
    179–85
Darwin, Charles 21–2
David, Ernst 222
DDT 92–3
de Grasse-Tyson, Neil 234
deficit model in science
    communication 210
defoliants 92
DeKalb Genetics Corporation
    33, 49–50, 80, 85, 86
Delta Pineland 65
Demeter movement 203–4
Denmark
    activism 266
    cassava 265
    labelling (in Copenhagen
        supermarkets) 194
    Sjömagasinet (restaurant)
        243–5
Department of Agriculture, US
    (USDA) 140, 148, 155,
    195–6, 203, 253
dermatitis and celery 192–3
derris dust (rotenone)
    199, 200
dhurrin 263
dicotyledonous plants 47–8
dioxins 992
disease resistance 68–73
    historical perspectives 22
    papaya 72, 119

potato late blight 218,
    225, 226, 227, 230,
    232, 285
tomato, transfer to
    potato 30–1, 39–41
wheat, transferred to
    sorghum 165
DLF Trifolium 91, 256
DNA 35–40, *see also* plasmids
    interrupted/interspersed
    repeats of, CRISPR
    and 257
    repair
        CRISPR and 249
        synthetic biology
            and 260
    transcription to RNA
        36, 58, 59
    transformation using *see*
        genetic transformation
double bonds in fatty acids
    144
Doudna, Jennifer (and her
    group) 256, 257, 258
Dow 75, 78, 80, 86, 88, 92
downstream regulatory
    sequences 38
downy mildew (on grapes)
    222, 223
Drenth, André 218–20, 221,
    222, 224, 225
drought
    maize tolerant to 280,
        281
    Southern Africa 267
DuPont 28, 48, 77, 80, 86, 88

Eastern brown snake, author
    encountering 104–5
e'cco Bistro 16, 17

eco-warriors *see* anti-GM activists and protestors
Edinburgh, Roslin Institute 291
egg allergy 18
Einstein, Albert 208
Elcano, Juan Sebastien 8
embryo rescue 72
emotions 1, 16
endosulfan 63, 65, 67, 106, 108
engineering part of synthetic biology 261
enhancer regions of promoters 58
5-enolylpyruvalshikimate-3-phosphate synthase (EPSPS) 56, 57, 58
environmental footprint of agriculture 278, 286
Environmental Protection Agency (EPA) 153, 154
enzymes
    herbicides unable to bind to isoforms of 57–8, 59
    synthetic biology and 261, 264–5
EPA (Environmental Protection Agency) 153, 154
EPSPS (5-enolylpyruvalshikimate-3-phosphate synthase) 56, 57, 58
Eriksson, Dennis 228–9
Escherichia coli 32, 39, 40, 127
ethephon 168, 169
ethylene
    pineapple 168, 169, 170
    tomato 160

Europe
    activism in *see* anti-GM activists
    biosecurity and biosafety *see* biosecurity and biosafety
    maize 115, 117–18
    potato late blight 216, 217, 220, 221
    starch market 229–30
European corn borer 68, 116, 117
European Food Safety Authority (EFSA) 230
European Patent Office
European Union (EU) 228
    potato and 228, 230
    regulators/regulation/legislation 90–1, 228, 230
"evil" corporations 92, 233

FAD2 (fatty acid desaturase 2) 146
fake news 137, 182
farming *see* agriculture
fatty acid 143–5, 146
FDA (Food and Drug Administration) 72, 153, 154, 159
fertilisers (chemical)
    historical perspectives 76–7
    nitrogenous *see* nitrogenous fertilisers
    opposition to use 176
    phosphate *vs* phosphite in 284–5
FiBL (Research Institute of Organic Agriculture) 237–41, 294, 295

fibre
    production 55, 276, 277,
        283, 291, 293
    synthetic 77
Finland, petunias 272–4
First World War in France,
    Australian soldiers 3–6
Fitt, Gary 66–7
FK290 (cotton variety) 101
Flanagan, Vincent George 3–5
flavour (taste) 12, *see also*
    smell
    super-tasters 13–14, 15
    sweetness 14–15
    vanilla 262–3, 266, 267
Flavr Savr tomatoes 55, 158,
    160, 161, 162
flower colours, petunias 273–6
flu *see* influenza
Folta, Kevin 236–7
Fontane potato 226
food 1–19
    allergies *see* allergies
    labelling 94–5, 161, 196,
        254, 275
    needs and uses and rela-
        tionships with 1–19
    organic *see* organic agri-
        culture and food
    pH 211
    prices *see* prices
    security 28, 62, 73, 131
Food and Drug Administration
    (FDA) 72, 153, 154, 159
Food Evolution 133–4, 135,
    137, 236
Fortuna potato 214–15,
    226, 230
Fourth Estate 282,
    see also media

Fox Paine 82, 83, 84
fragrant rice 252–7
Fraley, Robb 41, 46, 236
France
    activism 166, 228–9
    food in 10–11
    grape sprayed with copper
        sulphate in
        Bordeaux 222
    maize 118
    WW I Australian soldiers
        3–6
Frankenfoods 68, 149, 162
Frazer, Ian 94
freeze-drying of potato 225
Friends of the Earth 55, 151,
    167, 170, 227, 295
Friesians (Holsteins) 250,
    251, 252
fruit
    ripening *see* ripening
    softening 160
fungicides 126, 198, 218, 223,
    224
    potato light blight 218,
        223–4
furanocoumarins 191,
    192, 193
Fusarium 116, 118
    banana 118, 122–3,
        125, 127
future perspectives 270–99

Gao, Caixia (and her group)
    255–6, 266, 288
Gardasil 94
Garst 86, 150, 160–1
gene(s)
    basics about 35–42
    chimeric 37, 40, 46

editing 243–59, 290
  animals 291–2
  CRISPR 129, 243–59,
    291, 293
  expression 36–9, 58, 59,
    246
  LEGO block model 36–8,
    39, 40, 58
  silencing 160, 169,
    see also RNAi
gene gun and shotgun transfer
  48–9, 50, 72, 105
genetic diversity 197, 240
Genetic ID 151, 152
genetic technology, new 129,
  240, 243–69
genetic transformation 25, 26,
  30–3, 40, 41, 42, 43–6, 48,
  49, 51
  wheat 256
Genetically Engineered Food
  Alert 151, 152
Genus 291
Germany 53–4, 96–7, 163–4,
  166
  E. coli disease outbreak
    127
  maize 134
  protests/activism 53, 164,
    166, 167–8, 203, 215
Getaria (Spain) 8–9
Glickman, Dan 155
global warming *see* climate
  change
glucosides, cyanogenic
  263–5
glufosinate resistance 90,
  103, 150
gluten 19, 60
  gluten-free wheat 290

glyphosate 56, 57–8, 91, 196,
  199, 200–2
  soybean 109–10, 111–12,
    114, 201–2
Golden rice 98, 163, 231, 233,
  271, 295–6
Göttingen University 96–7
government regulation *see*
  legislation and regulation
grain *see* cereals
Grains Research and Development Corporation (GRDC)
  91
grape sprayed with copper
  sulphate 222
green parties and greens
  Australia 142
  Europe 164, 229
Green Revolution 29, 78, 98
greenhouse gas, methane
  as 286–8
Greenpeace 53, 55, 61, 112–13,
  113, 118, 131, 174, 231, 233,
  295, 296
Grierson, Don 160, 161
Griffith, Frederick 31
groundnut (peanut) allergy 18
Groupe Limagrain 85–6
growth hormone
  athletes 140
  bovine 148
guanine (G) 36
gun, gene (and shotgun transfer) 48–9, 50, 72, 105

H1N1 influenza 292
H5N1 influenza 292
Haber–Bosch process 20, 26,
  78, 176, 206
Hamill, John 34–5

Harding, Rob (and his group)
    68–71
Hawaii 130–6
    maize and soybean
        135, 136
    papaya 71–2, 73, 119,
        130, 130–6, 155
    sugarcane 130, 135, 136
health and safety issues 164,
    210, *see also* toxicity
    cassava 264
    EU 230
    fats 145
    GM pigs 147–9
    organic *vs* GM foods 195
Heisenberg's Uncertainty
    Principle 208
Helicoverpa (cotton boll
    weevil) 62, 64, 65, 66, 68,
    104, 108
Helicoverpa armigera (cotton
    bollworm) 62–3, 65
Helicoverpa punctigera (cotton
    budworm) 62–3
Henry, Robert 255
herbicides (weed control and
    weed killers) 109–15, 238
    maize 68
    phosphite as 286
    resistance and tolerance
        50, 56–61, 90, 96, 103,
        109–15, 116, 150, 184,
        185, 196
    soybean 59, 108–15,
        201–2
Herrera-Estrella, Luis 45, 47,
    283–4, 286
heterosis 25
Highland potato famine
    (Scotland) 216

Hitler and Jewish scientists 97
Holstein (Friesians) 250, 251,
    252
homeopathy 198, 204, 205
homology-dependent repair
    (HDR) and CRISPR 248, 249
Honeycutt, Zen 136–7, 142, 206
horn (cow/cattle)
    CRISPR-edited cattle
        without horns (hornless/
        polled) 250–2
    manure (preparation
        500) 201, 204–6
    silica (preparation 501)
        204, 206
Howard, Philip 84
HPV vaccine 94
human genome editing 257
Hurricane Irma 237
hybrid vigour 25

identity, food as part of 9–10
immune system, bacterial,
    CRISPR and 247
inbreeding 23–4, 24, 25
    depression 175
India 77–8
    Advanta in 83
    cotton 98–9, 175, 176,
        177–9
    farmer suicides 175–9
    wheat 28–9
indigo 77–8, 266, 268
Indonesian dish, pecel lele 9
Industrial Revolution 75, 278
Industrial Revolution 4.0 260,
    291
influenza 292–3
    Spanish 71, 292
Ingard cotton 65, 66

insect(s) (as pests)
  cotton, Australia 62–3,
    65–6, 67, 103–8
  resistance to 55, 62–8,
    101, 103, 192, 203, 240
insecticides
  cotton 98, 99, 100,
    106, 108
  DDT as 93
  spraying *see* spraying
Irish potato famine 21, 216
Irma (Hurricane) 237
isoflavones 14, 15
Italy, olive oil 147

Jannson, Stefan 244, 245
jasmine rice 252, 253, 254
jassid, cotton 107
Jewish scientists in Hitler's
  Germany 97
Johnson, Philip 16–17

Kahl, Gunter (and his group)
  48
kanamycin 40, 46
Kanzara (village in India) 98
Kelvin, Lord 209
Kennedy, Jonathon 178
Kennedy, Scott Hamilton
  133–4, 135, 137, 236
Kerala 177–8
King, Lawrence 178
Kinki University pigs 143–9
Kipling, Rudyard 11
kiwi fruit allergy 154
Kniss, Andrew 110
Kranthi, K.R. 99
Krattinger, Simon 165
Kuntz, Marcel 166, 167, 171
KWS 167–8, 227

labelling 94–5, 161, 196,
  254, 275
landscape changes, rural
  135–6
law *see* legislation and
  regulation
Lawes, John Bennet 76–7
LD_{50} *see* lethal dose 50
leaf blight, taro 69
leafhopper (jassid), cotton 107
legislation and regulation 164,
  233, 275–6
  EU 90–1, 228, 230
  Finland 274
  US 203
LEGO block model of
  genes 36–8, 39, 40, 58
legumes and paleo diet 138
Leibig, Justus von 76
Lemaux, Peggy 49–51, 195,
  236
lethal dose 50 (LD_{50}) 196, 197
  fungicides (for potato late
    blight) 223, 224
  glyphosate 202
  rotenone 199
  water 198
lettuce 151, 190, 191, 276
Liberty Link (glufosinate) resistance 90, 103, 150
Limburgerhof, BASF 213,
  214, 227
linamarin 264, 265
linoleic acid 146, 147
linolenic acid 144, 145, 154
lipids 143–5
lipoxygenases 15
Lobster, Claude 252
Losey, John 180, 182
Ludwigshafen 213, 214

lycopene 158
Lynas, Mark 230–3, 234,
    271, 295

McDonalds 206, 226, 228–9
MacKay family (Australia)
    128–9, 130
Madagascar, vanilla 263, 267
Madame Rouge
    (restaurant) 17
Magellan, Ferdinand 8
maize (corn) 117–19, 179–85,
    205, 280–1
    Bacillus thuringiensis see
        Bacillus thuringiensis
    drought-tolerant 280,
        281
    Germany 134
    Hawaii 135, 136
    herbicides 68
    historical perspectives 26
    hybrid vigour 25
    Mexico 179–85
    StarLink 149–55, 181–2
    United States 26, 115,
        117, 280–1
    Zambia 174
mammals, salt toxicity 197
mancozeb 223, 224
Mannheim 213, 214
Maori people 7
mating type A1/A2, *Phy-*
    tophthora infestans 218,
    220, 221, 226
Maurin, Jost 240
meat-less diet (vegetarian or
    vegan) 139, 146, 201, 254
media 166, 282, *see also* fake
    news
    social 137, 166, 210–11

Mediterranean corn borer
    (*Sesamia nonagrioides*) 118
Melanesia 8, 68, 69, 70, 71,
    199
Ménage a Trois (restaurant) 16
Mendel, Gregor 21, 22
methionine 139–40, 141, 142
Mexico
    Borlaug in 28
    Bt maize and Monarch
        butterflies 179–85
    Herrera-Estrella in 45–6,
        283
    potatoes and late blight in
        and from 215, 217,
        220
    Schilperoort in 45–6
Meyer, Peter 274
mildew
    downy (on grapes)
        222, 223
    resistance in wheat 290
milk
    school 7–8
    smell affected by cows
        ingesting a weed 190
milkweed and Monarch
    butterflies 180, 181, 182,
    184, 185
Millardet, Pierre 222
Møller, Birger Lindberg (and
    his group) 263, 265, 266
MON810 181, 182, 228
Monarch butterflies and Bt
    maize 179–85
monocotyledonous plants 48
monocultures, cotton 62,
    175
monounsaturated fatty acids/
    oils 144, 145

Monsanto 33, 46, 47, 50, 56, 58, 59, 61, 72, 75, 80, 83, 84–5, 86, 87–8, 90–4, 98, 234–5
  author's failed attempts to contact 235
  Bayer's takeover of 88, 90–1
  cotton 65, 66, 98, 99, 100, 101, 102, 103, 271
  fate 91
  maize 181
  protestors against 91–2, 168, 231, 234
  soybean 113–14
Montpelier, author in 10–11
Moreton Bay bugs 17
Müller, Paul 92–3
multinational corporations 133, 136, 163, 233–4, 234
Murray, Lester 13
mushroom, CRISPR-edited 244

national identity 9–10
nematode, root knot 39
neomycin-type antibiotics 40
Nestor's lab 42, 43
Netherlands, organic farming 219
new technologies 46–7, 98, 209, 234, 241, 293
  breeding (NBTs) 241, 244, 250, 297
  genetics 129–30, 240, 243–69
  regulation *see* legislation and regulation
New Zealand, biosecurity *see* biosecurity
NGOs 228, 241

n-hexanal 14, 15, 254
Nicotiana tabacum (tobacco) 34, 40, 46, 47, 72, 74–5, 140, 193–4, 285, 286
nicotine 74
Nigeria, cassava 265
Niggli, Urs 237, 238–9, 240, 241, 294, 295
nightshades *see Solanum*
nitrogenous fertilisers
  Haber–Bosch process 20, 26, 78, 176, 206
  opposition to use 176
  preparation 500 as 204–5
non-government organisations (NGOs) 228, 241
non-homologous end joining (NHEJ) and CRISPR 248
nucleotides 36
  in DNA repair 249
nut allergies
  groundnut (peanut) 18
  tree nut 140, 141
nutritional value of soybean 138–42

Øcological (on label) 194
octopine 44
oils (dietary) 143–7
  olive oil 135, 143, 144, 145, 147
Öko (on label) 194
oleic acid 144, 145, 147
olive oil 135, 143, 144, 145, 147
omega-3 fatty acids 144
oomycetes 285
  Phytophthora infestans 215, 217, 218, 220, 285
opines 44

Organic 3.0 239
organic agriculture and
    food 14, 173–4, 186–7, 195,
    203–4, 237–41, 296–7, *see
    also* biodynamic agriculture
    allowable inputs and
        LD$_{50}$ 197–8
    certification 186, 187,
        193, 195, 196, 197, 199,
        200, 203, 239, 240, 241,
        279, 293, 297
    conventional agriculture
        and
            binary divide
                276, 297
            combination of 283,
                297
    Netherlands 219
    restrictive practices 297
    social media and
        210–11
    Switzerland 237–41
Organic Agriculture Research
    Institute (FiBL) 237–41,
    294, 295
Organic Consumers
    Association 151, 236
orthophosphate 284, 285
Ostrinia nubialis (European
    corn borer) 68, 116, 117
outcrossing 24
Oxford Farming Conference
    (2013), Mark Lynas
    at 230–1, 235

Pacific islands/nations *see*
    Hawaii; Melanesia;
    Philippines; Polynesia;
    South Pacific
Pacific Seeds 82, 83

Pakistani, basmati rice *vs*
    jasmine rice 252
paleo diet 138–9
palmitic acid 143, 145, 146,
    147, 254
Paltrow, Gwyneth 206
PAM and CRISPR 248
Panama disease 122–9, 155
papaya (*Carica papaya*)
    Hawaiian 71–2, 73, 119,
        130–6, 155
    Rainbow 71–3, 72, 119
papaya, Philippines and
    Thailand 119, 130–1
papaya ringspot virus
    (PRSV) 71, 72, 73, 130–1
    resistance 71–2, 119
Papua New Guinea 69, 71, 173,
    221–3
Paraguay, soybean 112, 114
Pascal celery 192
patents
    CRISPR 257–9
    glyphosate 91, 114
PCR (polymerase chain re-
    action) 70–1, 152, 153
peanut allergy 18
pecel lele (Indonesian dish)
    9
pectin 159, 160
personal identity 9, 10
pesticides, GMOs as 185–6, *see
    also* fungicides; herbicides;
    insecticides
pets, chocolate toxicity 197
petunias 47, 59, 272–6
pH of food 211
Philippines
    papaya 119, 130–1
    rice 252

phosphite and phosphate 284–6
photosynthesis 285
    crassulacean acid metab-
        olism (CAM) 169–70
    herbicides interfering
        with 56, 59
Phytophthora colocasiae 69
Phytophthora infestans see
    potato late blight
pigs 142, 143–9, 155, 291–2
    porcine reproductive and
        respiratory syndrome
        291–2
pineapples
    Hawaii 135
    Queensland 168–70
Pioneer Hi-Bred 26, 50, 80, 86,
    235, 280
Pisang Mas 123
plant(s), *see also* crops;
    herbicides
    callus tissue 46, 49
    cell wall 33, 34, 159
    cotyledons 47–8
    defoliants 92
    disease resistance *see*
        disease resistance
    fats/lipids 144, 145
    genetic transformation *see*
        genetic transformation
    photosynthesis *see*
        photosynthesis
    protoplasts 33, 34, 35, 40
plant breeders 20, 21, 22, 29,
    72, 226
plasmids 31–2, 40
    A. faciens (Ti) 42, 43, 44, 48
    in protoplasts 35
Plasmopara viticola (downy
    mildew) grapes 222, 223

plum jam/plum and apple jam
    and WWI 3–6
polled (hornless) cattle,
    CRISPR-edited 250–2
pollen and pollination 24, 25
    Bt maize pollen 179, 180,
        181–2, 182
polymerase chain reaction
    (PCR) 70–1, 152, 153
Polynesia 68, 71
polypeptide synthesis from
    RNA *see* translation
polyunsaturated fatty
    acids 144, 145, 149
poor people 278–9
porcine reproductive and re-
    spiratory syndrome 291
pork 143, 146–7, 149, 291
post-translational modifi-
    cations 58–9
potassium in bananas 60
potato (*Solanum tuberosum*)
    214–27, 229–30
    Amflora 228–9
    CRISPR-edited 244
    Fontane 226
    Fortuna 214–15, 226, 230
    fusing cells of tomato
        and 35
    Sequoia 222
    tomato disease resistance
        transfer to 30–1, 39–41
potato late blight (*Phytophthora
    infestans*) 21, 22, 215–17,
    219–26, 230, 232
    resistance to 218, 225,
        226, 227, 230, 232, 285
Potrykus, Ingo (and his group)
    47, 50, 163
poultry birds, flu 292, 293

poverty 278–9
preparation 500 (cow horn
    manure) 201, 204–6
preparation 501 (cow horn
    silica) 204, 206
prices for agricultural products
    (incl. food) over time 87
    price-fixing deals 317
productivity (improving)
    1980s and companies
        involved in 280
    subsistence farming 280
profit-making by companies
    75
pro-GM activists 163, 236
promoter enhancer regions 58
propamocarb 223, 224
protein
    synthesis from RNA see
        translation
    transport to chloroplast 59
protestors see anti-GM activists
    and protestors
protoplasts 33, 34, 35, 40
protospacer adjacent motif
    (PAM) and CRISPR 248
PRSV (papaya ringspot virus)
    71, 72, 73, 130–1
Pseudomonas
    P. syringae 283
    phosphite oxidoreductase
        283
psoralens 191, 193
public (general) 49, 154, 161,
    185, 209, 294, 295
    communication to
        49, 209
    interests of 283–4
Puccinia see rust
pulses 139, 140

Qaim, Martin 53, 96, 97–8,
    112, 115, 238
Queensland
    agricultural changes from
        1960s to now 135
    Agricultural College 77
    banana and Panama dis-
        ease 126, 127, 128, 129
    celery 191–2
    choco and school milk
        7–8
    papaya 130
    pineapples 168–70
Queensland Alliance for
    Agriculture and Food
    Innovation 255

Rainbow papaya 72, 119
Reblo (papaya) 130
Recombinetics 252
regulatory bodies see legisla-
    tion and regulation
regulatory sequences on DNA/
    genes 38
renewables vs coal in
    Australia 207
Research Institute of Organic
    Agriculture (FiBL) 237–41,
    294, 295
resistance (and tolerance)
    antibiotic 31–2, 40, 46, 49
    disease see disease
        resistance
    herbicide 50, 56–61, 90,
        96, 103, 109–15, 116,
        150, 184, 185, 196
    insect 55, 62–8, 101, 103,
        192, 203, 240
restrictive practices in organic
    farming 297

revolutions (in agriculture and food production) 30, 52–73
  in agbiotech *see* agricultural biotechnology
  Green Revolution 29, 78, 98
rice 252–7, 286–8
  fragrant 252–7
  Golden 98, 163, 231, 233, 271, 295–6
  methane and 286–8
Right To Know (RTK) 236
Rio Grande do Sul (Brazil) 61, 113, 114–15
ripening of fruit (and its delay) 168
  tomato 55, 158, 159–60, 160, 161
RNA
  DNA transcription to 36, 58, 59
  single guide (sgRNA), CRISPR and 248
  translation *see* translation
RNAi (RNA interference) 160, 265
Romeis, Jörg 165
Ronald, Pam 283
root knot nematode 39
Roslin Institute 291
rotenone 199, 200
Roundup and Roundup Ready 56–61, 69, 90, 103
  soybean 59, 108–15
Rpi-blb2 and Rpi-blb2 genes 226
RuBisCO 59
rural landscape changes 135–6

rust (*Puccinia*)
  maize 122
  wheat, resistance transferred to sorghum 165
Ryan, Cami 94

Sacrificial Virgins 93–4
Saedler, Heinz 274
safety *see* biosecurity and biosafety; health and safety; toxicity
Safeway 158, 161
Sainsbury's 158, 161, 162
salt (sodium) toxicity 196, 197, 198–9
same-sex marriage in Australia 207
Sanford, John 48, 72
saturated fats 143, 144, 145
Schell, Jeff (and his group) 33, 41, 42, 45, 46, 283
Schilperoort, Rob 45
science 209–10
  creationist scientists 207
  values and 206, 207
Scottish Highland potato famine 216
seafood and shellfish 13, 17–18
  allergy 18
Sechium edule (choko or chayote) 6–7
seed
  companies 33, 79, 80–1, 81–2, 84, 99–100, 135, 150, 167–8, 210, 227, 274, 282
    largest 85
  Hawaii as base for production 135–6
Sequoia potato 222

Sesamia nonagrioides 118
sexual incompatibility, plant
    species 31
sexual reproduction, *Phy-*
    tophthora infestans 218, 220
shellfish *see* seafood and
    shellfish
shills 133, 137, 235, 236, 237
Shiva, Vandana 175, 177, 233
shotgun transfer (gene gun)
    48–9, 50, 72, 105
Sigatoka diseases 121, 126,
    127, 128
Silent Spring 92
single guide RNA (sgRNA),
    CRISPR and 248
Sjömagasinet (restaurant) 243–5
skin prick test for allergy 140–1
skin reaction to celery 192–3
Slighton, Jerry 72
smell/odour/aroma 106
        milk affected by cows in-
            gesting a weed 190
        rice 252
Smith, Jeffrey 133–4, 152
Smyth, Stuart 276–7, 294,
    295, 296
Smythe, Heather 12–14, 16
snake, author encountering
    104–5
social media 137, 166, 210–11
sodium (salt) toxicity 196, 197,
    198–9
softening of fruit 160
soil tilling, soybean 111
Solanum (nightshades)
        S. belladonna 30
        S. bulbocastum 226
        S. lycopersici see tomato
        S. tuberosum see potato

Somme (and WWI battle) 2,
    3, 5, 190
sorghum 66–7, 80, 83–4, 104,
    165, 185–6
        cyanogenic glucosides
            263, 264
        wheat rust resistance
            gene transferred to 165
South America, soybean 61,
    109, 112–15
South Pacific, *see also*
    Melanesia; Polynesia
        papaya 72–3, 119
        taro 68–9
South West corn borer 117
Southern Africa, drought
    267
soy/soybean 60–1, 108–15,
    138–42, 155, 201–2
        allergy 140–1, 155
        Hawaii 135, 136
        herbicides incl. Roundup
            Ready 59, 108–15,
            201–2
        nutritional value 138–42
        products and chemicals
            14, 15
        trypsin inhibitors 60,
            202
Spain
        Getaria in 8–9
        maize 117–18
Spanish influenza 71, 292
spongiform papillae on
    tongue 13–14
spores
        Panama disease (banana)
            125, 127
        potato blight 217, 218,
            220, 221, 222

spraying with insecticide 203
    Australia 62–3, 65, 67,
        105–6, 108, 126, 136
    Burkina Faso 271–2
    France, grapes 222
    Spain 117
starch
    European market 229–30
    rice 288
StarLink maize/corn 149–55,
    181–2
stearic acid 143, 144, 145, 147
Steiner, Rudolf 203
subsistence farming 2, 278,
    279–82, 295, 296
sugar signalling in barley A2
    gene 288
sugarcane 130, 135, 136
    Hawaii 130, 135, 136
Sun, Chuanxin (and his group)
    288
sun exposure and psoralens
    191–2, *see also* ultraviolet
SunRice 253, 254
super-tasters 13–14, 15
supermarket tomatoes 55,
    158, 160, 161, 162
superphosphate 76–7
SUSIBA2 gene, barley 288
Swaminathan, M. S. 28, 29
sweetness 14–15
    artificial sweetener 93
Switzerland 54, 164–6
    organic farming 237–41
Syngenta 82, 86, 87–8, 117
synthetic biology 259–68

2,4,5T 92
TALENs 250, 256
Tanty, Francois 11

Tanzania 279–80, 281
tapioca (cassava) 264–5
taro (cocoyam) and aroid root
    crops 68–70, 71
taste *see* flavour
taste buds 9, 15
Teeri, Teemu 273–4
terroir 12, 13, 263
Thailand
    jasmine rice 252, 253
    papaya 119, 130–1
Thenus orientalis 17
theobromine 197, 198, 199
35S promotor 58
thymine (T) 36
Ti (tumour-inducing) plasmid
    (*A. faciens*) 42, 43, 44, 48
Ticklers plum (and apple) jam 5
tillage of soybean 111
tobacco (*Nicotiana tabacum*) 34,
    40, 46, 47, 74–5, 140, 193–4
    cancer and 193–4
    grown on phosphate *vs*
        phosphite 285, 286
    resistance to tobacco
        mosaic virus 72
tofu 14, 15, 201
tolerance *see* resistance (and
    tolerance)
tomato (*Solanum lycopersici*)
    30, 35, 158–62
    disease resistance, transfer
        to potato 30–1, 39–41
    Flavr Savr 55, 158, 160,
        161, 162
    fusing cells of potato
        and 35
    ripening (and its delay)
        55, 158, 159–60,
        160, 161

tongue
    spongiform papillae
        on 13–14
    taste buds 9, 15
Torbitt, James 22
toxicity (adverse health
    impact) 173–5, *see also*
    lethal dose 50
    cassava 264
    studies (toxicology)
        196–9
            potato late blight-
            controlling chem-
            icals 223–4
toxins
    Bacillus thuringiensis see
    Bacillus thuringiensis
    food crops/edible plants
        containing 60
    herbicides as 57–8
transcription (DNA to RNA)
    36, 58, 59
transcription activator-like ef-
    fector nucleases (TALENs)
    250, 256
transformation 29
    genetic *see* genetic
        transformation
transit peptide 58–9
translation (RNA to protein/
    polypeptide) 36–7, 58–9
    modifications following
        58–9
tree nut allergies 140, 141
tribalism 172–3
Trifolium (DLF Trifolium) 91,
    256
Tropical Race 4 (TR4) Panama
    disease 123–4, 126, 127,
    128, 129

trypsin inhibitors, soybean
    60, 202
2S albumin 140, 141

UCP1 291
ultraviolet (UV) light, *see also*
    sun exposure
    DNA and 249
    psoralens and 191–2
Uncertainty Principle
    (Heisenberg's) 208
uncoupling protein 1 291
United Phosphates Limited 83
United States (USA)
    Centers for Disease
        Control and Prevention
        (CDC) 142, 154
    Department of Agri-
        culture (USDA) 140,
        148, 155, 195–6,
        203, 253
    Environmental Protection
        Agency (EPA) 150,
        153, 154
    food allergies 18, 18–19
    Food and Drug Adminis-
        tration (FDA) 72, 153,
        154, 159
    maize 26, 115, 117,
        280–1
    regulation 203
    soybean 60–1, 61, 109,
        112
University of California,
    Berkeley, CRISPR 256, 257,
    258, 259
unsaturated fats 144, 145, 149
upstream regulatory
    sequences 38
UV *see* ultraviolet

vaccine to HPV 94
values (human) 172–3, 193, 194, 196, 200, 205, 206–9
van Eenennaam, Alison 244, 246, 250
van Montagu, Marc (and his group) 33, 41–2, 44, 46, 86, 283
vanilla 261–3, 266–8
vanillin 262, 263, 266
Vegemite 9
vegetarian or vegan diet 139, 146, 201, 254
virus resistance 69–73
  papaya 71–2, 119
Visy and Amcor 89–90
vitamin A and Golden Rice 271, 296
von Baeyer, Adolf 78
von Leibig, Justus 76
Voytas, Dan (and his group) 288, 290

Wallace, Henry A. 26–8
Wee Waa 104, 106
weed control and weed killers *see* herbicides

wheat
  gene editing 290
  historical perspectives 28–9
  nitrogenous fertilisers 28
  rust resistance gene transferred to sorghum 165
whiteflies 99, 108
wholegrain cereals 139
Wichtrich, John 152–3
World War I in France, Australian soldiers 3–6

xylem, banana tree 124–5

yeast, vanilla biosynthesis 266
yellow Sigatoka 121, 126

Zalucki, Myron 184
Zambian maize 174
Zambryski, Patricia 44, 45, 46
Zeneca 160, 162
Zhang, Feng (and the Broad Institute) 257–8, 259
Zhu, Jian-Kang 288
zinc finger nucleases (ZFNs) 250, 256